Animal Behavior
A Beginner's Guide

John A. Byers

ONEWORLD

A Oneworld Paperback Original

First published by Oneworld Publications, 2013

ISBN 978-1-78074-260-1
eISBN 978-1-78074-261-8

Typeset by Cenveo Publishing Services
Printed and bound in Great Britain by
TJ International, Padstow, Cornwall, UK

Oneworld Publications
10 Bloomsbury Street
London WC1B 3SR
England

Contents

Preface

The goal of this Beginner's Guide is to show that the study of animal behavior is an integrative discipline. What we recognize as behavior is the outcome of patterns of muscle contraction. Muscle contractions pull on the skeleton, and movements result. All the motions in a species' repertoire are repeating units; they can be described and cataloged. These movements carry animals through their environments, harvest food, produce signals to other species members, transfer gametes, build things, and care for offspring. Muscles are directed to contract by the nervous system, which integrates information from the sense organs and internal sensors, and makes moment-to-moment decisions on which actions the musculoskeletal system should produce. We can study how these moment-to-moment decisions contribute to individual survival and reproductive success, and we can study the nervous system mechanisms that direct the moment-to-moment switching of motor output. To fully understand behavior, we need to understand evolutionary theory, ecology, population biology, genetics, physiology, anatomy, and nervous system design and function. The modern study of animal behavior seeks to integrate these spheres of knowledge into a comprehensive understanding of how and why animals move in the elegant ways that they do to reproduce successfully within their social and environmental worlds. Scientists pursue this mission in the laboratory, where strict control of the environment is needed, in the backyard, park, or urban environment, where small animals can be completely at home, and in the wilderness, which is the only place to observe some of the world's most beautiful animals in what is home to them: an unaltered natural environment.

1

The Biology of Behavior

Imagine that you are sitting at your kitchen table. It is a beautiful summer morning; the door slaps shut after the dog has pushed it open with her nose to go outside. As you take your first sip of coffee, a housefly that entered when the dog exited suddenly claims your attention. Like a tiny vulture, the fly circles above the table, slowly descending, until she lands close to the sugar bowl. The fly walks toward the bowl, and stops by a few grains of sugar that you spilled when you lifted your spoon from the bowl two minutes earlier. The fly inflates her proboscis and begins to dab at the sugar. As you watch this, you begin to feel a mild sense of outrage, not because the fly is stealing sugar, but because the fly's moist, spongy proboscis, now dabbing at your table, was recently outside, probably dabbing at dog feces or the rotting chicken in the rubbish. You wave your free hand at the fly; she jumps into the air and hovers nearby before quickly landing back at the sugar. You bring your hand rapidly down, attempting to crush the fly, but she is too quick for you. You put your coffee cup down, rise, and reach for the flyswatter, a tool that humans, with their big brains, have invented to crush flies. The flyswatter doubles the effective length of your forearm and so doubles the speed of your strike. The fly has resumed dabbing at the sugar. You strike. The fly sees the rapidly approaching head of the flyswatter. She jumps and begins to fly, but the broad head of your simple tool stops her flight and smacks her with enormous force into the table. Her internal organs, including her brain, are crushed beyond repair. A tiny marvel of miniaturized circuitry and engineering lies mangled on your table. In your own brain, the circuits that would trigger shame or remorse do not light up. You brush the carcass to the floor and step toward the door, where the dog is scratching to be let in.

The first thing to say about this ordinary moment in life is to note the extraordinary performances of you, the fly, and the dog. The three of you used sensors that are tuned to radiation in certain parts of the electromagnetic spectrum, as well as sensors (in the dog and the fly) tuned to detect certain chemicals in the environment, to extract useful information about your environments. Your brains then issued precise sets of commands to muscles that pulled in a complex way on your skeletons, causing your bodies or parts of your bodies to move through space in a smooth, goal directed way.

W.T. Thach, a neurobiologist who studies the cerebellum, a part of the brain that controls movements, once remarked that, "Moving the skeleton is an engineer's nightmare." Yet animals, such as you, the dog, and the fly, make smooth, precisely timed and impeccably directed movements. The way that each of you produces movements is essentially identical. You, the dog, and the fly have sense organs that transduce environmental energy or materials into patterns of nerve signals. Each of you uses sense organs, called eyes, to transduce light energy from a narrow band of the electromagnetic spectrum. Each of you has sensors that bind to certain chemicals in the environment and transduce this event into patterns of nerve signals. For you and the dog, these chemical sensors are in the nose; in the fly, the sensors are on many parts of the outer body, including the feet – that's how the fly knew to stop and extend her proboscis when she walked into that sweet spot on your table.

Each animal species also has a rich variety of internal sensors. You and the dog have receptors that report on blood temperature, sugar level, and acidity, the amount that each muscle is stretched, the likelihood that damage is occurring, mechanical pressure at most parts of the skin, how much the muscular walls of each blood vessel are contracted, and the position of your head and eyes, to list a few examples. The internal sensors report on the biochemical and mechanical integrity of the body, as well as on body part position. All these reports, from the sense organs and from the internal sensors go, in you, the dog, and the fly, to the brain.

A brain is an integration and command center. It receives reports from the sense organs and the internal sensors, integrates the information to create priorities and then, based on the priority list, issues commands. These commands are of two sorts. First, there are commands to the organs – to the heart, lungs, gut, blood vessels, and endocrine glands. These commands are concerned with the essential task of keeping the body running. However, as we saw in our summer tableau, an animal must do more than simply rest, plant-like, in a steady state; it must also move through its environment. This is where the second sort of command comes in. These are the commands to the muscles that pull on the skeleton and produce movements. These movements are known as behavior.

Muscle tissue is evolutionarily ancient and operates the same way in all animals. Commands from the brain, traveling along a motor nerve, reach a muscle. The muscle, in response to the commands, uses stored energy to contract: to become shorter. When the muscle shortens, one part of the skeleton moves with respect to another, and the animal moves in some way. A cheetah sprinting in pursuit of a gazelle, a bumblebee flying over a meadow to land on a flower, a great blue heron stabbing at a fish, bullfrogs calling, crickets chirping, Hillary Hahn playing the Bach D minor *Ciaconna*; all are the outcomes of patterns of muscle contraction.

We have only a dim understanding of the complex mechanisms by which a brain decides which commands to issue from moment to moment. However, biologists are certain about the design features that they expect – the general decision-making rules that brains use. Behavioral biologists assume that brains are organized so that from moment to moment, individual animals act as if they are asking themselves the following question: "What should I be doing at this moment to maximize my lifetime reproductive success?" In the past twenty years we have learned that animals in nature really do act this way. But to accept this premise, must we conclude that animals consciously make decisions and are capable of the mathematics, or at least the complex reasoning, required?

Clever Hans

In the late nineteenth century, Wilhelm von Osten startled the public and professional psychologists by showing that his Arabian stallion, Hans, could perform arithmetical calculations, including addition, subtraction, multiplication, and division using integers or fractions, take simple square roots, tell time, read, and spell. Von Osten put questions to Hans either orally or using a blackboard. Hans repeatedly pawed with a forefoot to indicate the correct answer. Von Osten: "Hans, what is the answer to 10 divided by 2?" Hans paws five times. Von Osten was convinced that animals have mental capacities that are essentially equivalent to those of humans, and he toured with Hans to prove this point. A panel of experts convened and concluded that no trickery was involved.

Then a student, Oskar Pfungst, arranged for questions to be put to Hans when neither the questioner, nor anyone that Hans could see, knew the answer to the question. In these trials, Hans could not provide the correct answer. Pfungst noted that when Hans, pawing with a forefoot, approached the correct answer, von Osten and others performed very slight, almost unnoticeable movements, such as raising or lowering the head by a fraction of an inch, or making tiny changes in facial expression. At the moment that Hans reached the correct answer, the release of anticipation in the human created a tiny twitch or jerk. What Hans had really learned was not math, but that in order to get von Osten's training rewards, he should start pawing in response to the postural cues that accompanied a question, slow his pawing when subtle postural changes indicated that he was near the answer, and stop pawing when he observed the little movement that indicated release of tension in the human. Pfungst had discovered the phenomenon of unintentional cueing, which modern experimental psychologists call the Clever Hans Effect, and assiduously avoid in their experimental designs.

There are several lessons embodied in the story of Clever Hans. One, relevant to this chapter, is that animals can act as if they have complex contingencies in mind, and have conscious goals. Hans acted as if he could perform arithmetic, when the mechanism underlying his performance was simpler. A housefly can act as if she is constantly thinking about what she should be doing from moment to moment to maximize the lifetime number of her eggs that hatch, when her brain is far too small to produce anything like conscious thought.

Ethograms

In 1973, the Nobel Prize in Physiology or Medicine went to three European biologists: Konrad Lorenz, Niko Tinbergen, and Karl von Frisch. These men received the prize not only for the significant insights and discoveries that each had contributed, but also because they were leaders in a new field of biology: ethology. The Nobel Prize affirmed that ethology represented a significant intellectual shift – one might even say a thought revolution – in biology. What was the nature of the shift? It was that the behavior of animals was a proper subject for biological study. In other words, just as one may study a species' skeleton, or organs, or muscles, so one may study a species' behavior. There is a biology of behavior.

The thought revolution came from a deceptively simple technique – watching animals – but watching animals for long periods of time with patience, unwavering focus, and a willingness to suspend interpretation. When you watch an individual animal in this way, you begin to see that behavior is not continuous improvisation. There are repeating units.

The ethologists referred to these repeating units as Fixed Action Patterns. The designation "fixed" meant that the movement in question had the same form every time that it was performed. "Same form" means two things. First, if we describe the movement using technology such as slow motion video analysis, we see that the duration of the action, the sequence of flexion or extension of joints, and the degree to which each joint is flexed or extended, is identical every time that the action is performed. Second, and more fundamentally, the sequence and duration of individual muscle contractions is the same each time. We can discover whether this is so using a physiological technique called electromyography (EMG). When a muscle contracts, a weak electrical event sweeps across its surface. By attaching electrodes to the muscle, we can monitor when it is contracting and how forcefully it is contracting.

Any motion that you or any other animal performs is the result of a closely coordinated sequence of contractions of individual muscles. For example, just before you noticed the fly descending to your table, you raised the coffee cup to your lips, tilted it, and introduced some coffee into your mouth. Then, quite unconsciously, you moved the liquid in your mouth to the back of your throat, across the top of your windpipe, and into the top of your esophagus – you swallowed. That motion, swallowing, involved a precisely timed sequence of contraction of about a dozen separate muscles in your throat. Had one of the muscles started to contract one-tenth of a second too early or too late, the smooth movement of coffee into your esophagus would not have occurred.

Consider another famous Fixed Action Pattern made famous by ethologists – egg rolling. This motor pattern exists in a number of waterfowl species. Picture a goose parent, incubating eggs on a nest. Now we place an egg or an egg-shaped object in front of the nest. The parent may have been looking to the side, but it suddenly turns to stare directly at the egg. For the moment, the egg claims all the parent's attention. After a few seconds, the parent stretches its neck forward, places its bill over the far side of the egg, and tucks its bill toward its chest, drawing the egg into the nest. The action obviously evolved because parents that retrieved wayward eggs back into the nest left more surviving offspring than parents that did not. A variety of objects will suffice to initiate this motor sequence in an incubating parent, and once initiated, the action runs, machinelike, to its conclusion. One can even reach in and remove the egg – the parent will continue with the tucking motion as if the egg were still present.

The extraordinary feat of the ethologists was that they did not reach their conclusion about repeating units of behavior with slow motion analysis or EMG. They simply watched animals with enough attention to perceive this truth. They also realized that one could compile a catalog of all the action patterns in the repertoire of a species. The term that emerged for such a catalog was "ethogram." An ethogram for a species comprises a short descriptive phrase to denote each action, accompanied by a succinct description of the movement, for all movements performed by the species.

As ethologists continued to construct and to refine ethograms, they realized that not all motor patterns were as lengthy as egg rolling. However, the longer duration motor patterns were an entry point that allowed ethologists to appreciate that behavior is composed of fixed, repeating units. This fundamental insight has held up for over half a century. Today, there is some debate over the nature of the indivisible units, but little debate over their existence.

Another fundamental insight emerged from the construction of carefully conducted ethograms in an increasing number of species. This was that patterns of similarity and difference in ethograms mirror evolutionary relatedness. If two species have recently diverged from a common ancestor, the lists of motor patterns of the two species will be similar and the individual motor patterns will be similar in form. For example, humans and chimpanzees are closely related and they have similar ethograms. Both species laugh. Both beg by holding out a hand, palm up. In contrast, if two species are not closely related, then their ethograms are dissimilar.

That patterns of similarity and difference in ethograms match patterns of phylogeny (evolutionary relatedness) was powerful evidence that the ethologists were correct in their assertion that behavior is part of a species' biology. It was already widely known, from Darwin and earlier biologists, that observed patterns of anatomical similarity and difference corresponded to phylogeny. The ethologists of the first half of the twentieth century figured out a way to describe the "anatomy" of behavior and found the same pattern. Motor patterns are gained, lost, and modified within any evolutionary lineage of animals.

Beyond Ethograms – Tinbergen's Four Questions

In a now famous 1963 paper, Niko Tinbergen, who would share the Nobel Prize ten years later, outlined his thoughts on how ethologists should set research goals after they had learned to describe behavior using ethograms. Tinbergen had two main goals for his essay. First, he wanted to give full credit to Konrad Lorenz (another of the 1973 Nobel Laureates), who he considered to be the father of modern ethology. Lorenz, more than anyone else, should be recognized as the principal champion of the view that behavioral traits were biological traits. Second, Tinbergen argued that to fully understand any motor pattern, one needed to amass knowledge about it in four domains. We now refer to these as Tinbergen's four questions. The domains, in the order listed by Tinbergen, are:

Causation

This domain embodies two sorts of information. First, what external stimuli are necessary and sufficient to elicit the motor pattern? For an incubating goose, we know that the stimulus needed to elicit egg rolling must be essentially egg shaped, and we know that a larger than normal egg-like object is even more effective than a true egg. By experimentally altering the nature of the "egg" that we present to the goose, we can discover the precise aspects of a stimulus (the ethologists used the term "sign stimulus") that turn on (the ethologists used the term "release") the egg-rolling motor pattern. Second, what is the nature of the "wiring diagram," the brain architecture, that identifies the stimulus, makes the decision to turn on egg rolling, and issues the specific commands to muscles? For all but the simplest behavioral acts, we have only vague answers to this set of questions. Even for a housefly, the switching mechanisms that direct movement from one moment to the next are hideously complicated.

Survival Value

Next, Tinbergen turned to a domain that was controversial and little studied at the time: how a behavioral act contributes to the survival, and ultimately to the reproductive success of the animal that performs the act. When Tinbergen wrote his paper, many biologists thought that study of the function of behavior (how behavior contributes to survival and reproductive success) was not possible. Tinbergen pointed out that patient observation of animals in nature leads to hypotheses about function.

> First, as one becomes better acquainted with a species, one notices more and more aspects with a possible survival value. It took me ten years of observation to realize that the removal of the empty eggshell after hatching, which I had known all along the black-headed gulls to do, might have a definite function… (Tinbergen, 1963, p. 422.)

Further, Tinbergen showed, in pioneering studies, that experimental tests of these hypotheses could be conducted in nature. He tested the hypothesis that eggshell removal by gull parents contributed to the survival of young by scattering gull eggs in an area where bird predators of eggs hunted. He placed empty eggshells near some of the eggs, and showed that predators found these more quickly than the eggs with no eggshells nearby. Field tests of hypotheses about the function of behavior constitute the majority of research done in animal behavior today.

Ontogeny

Ontogeny means development. Here, Tinbergen applied his characteristic lucidity to a vexatious issue – the nature–nurture debate. In the 1960s, this was essentially a European–American debate. The European ethologists, with a tradition of observing animals in nature, contended that many behavioral acts, such as egg rolling, were instinctive, or "innate." The acts appeared in perfect form the first time that the animal perceived the appropriate stimulus. American psychologists, in contrast, had a rich tradition of experimentation and careful consideration of experimental design, using a few domestic species (mainly Norway rats, house mice, and pigeons) in controlled laboratory settings. The Americans contended that many, if not all, behavioral acts were influenced by learning. Thus, a contentious transatlantic debate arose on whether motor patterns were instinctive or learned.

Tinbergen pointed out that the debate as framed was not useful. He noted that labeling an act as either instinctive or learned is to make a claim about the role of experience in the development of the act. Labeling an act as instinctive implies that the act develops with no experience of any sort required. But, Tinbergen argued, proving that no experience is required means that we must experimentally prove that each independent possible source of experience is not needed: ultimately, we are faced with an almost endless number of negative proofs. A complementary difficulty arises when we want to claim that an act is learned. Tinbergen argued that, rather than labeling an act as instinctive or learned, and then quarrelling about it, we can more profitably just study the development of the act, recognizing that most acts will develop through a blend of innate and learned mechanisms.

Although Tinbergen, and many others since, pointed out that the debate was vapid, nature–nurture just won't go away: it arises, Phoenix-like, in the scientific literature and it is promulgated vigorously by the popular media, who are practiced in the preservation of simplistic controversies.

Evolution

Tinbergen's fourth and final domain was about evolutionary change in behavior. What is the pattern of evolutionary change in behavior, and by what process does such change come about? Ethologists had already shown that the more closely related two species were, the more similar were their ethograms. Therefore, the differences in ethograms among species must be the result of evolutionary change. Tinbergen pointed out that a careful comparison of ethograms across an array of related species would allow us to describe the pattern of evolutionary change in behavior. Consider a recent example: if we tickle a human or a great ape (chimpanzee, bonobo, gorilla, orangutan), one motor pattern that we elicit is laughter. Although the motor pattern is obviously recognizable as laughter in the five species, there are clear differences. When researchers constructed a family tree of the five species based on measurable aspects of the laugh sounds, they arrived at the same tree that one gets by comparing anatomy, amino acid sequences of proteins, or nucleotide sequences of DNA. Orangutans branch off first, then gorillas, leaving humans, chimpanzees, and bonobos clustered together, having and only recently diverged. We can also trace evolutionary change in the laugh sounds. For example, when humans laugh, there is repeated voicing on exhalation followed by silent inhalation: HA-breathe in HA-breathe in HA-breathe in HA. As chimpanzees diverged from the ancestors of humans, they increased the tendency to voice on inhalation and exhalation. Try it: say HA as you breathe out and as you breathe in: you will sound like a chimpanzee being tickled.

Tinbergen also noted that if behavior has, in his words, "survival value," meaning that it is specialized to fit the animal to its specific environment, then the behavior must have reached its current form through evolution by natural selection. If behavior is to change under natural selection, then: 1) individuals must vary in behavior, 2) this variation must create differences in reproductive success, and 3) the variation must in part be due to genetic differences among individuals that can be reliably inherited. Tinbergen argued that we needed to study these things. How does the genetic control of behavior work? How is behavior inherited? How rapidly can behavioral change occur when there is selection?

Tinbergen's Four Questions were a remarkably complete prescription for the study of behavior, and they also provided an enduring model of organized thought in biology. His thinking should guide us as we make our way through this book.

2
Multiple Realities

For millennia, philosophers mused about the nature of reality and consciousness. They sought to discover the truth simply by thinking about it. They made slow headway, but theirs was the only game in town. Then, in the nineteenth and twentieth centuries, empirical science amassed enough knowledge to provide another view. That view is that there is a single physical reality in the universe. Animal life, in its inexorable trend to become more competitive, more efficient, better at harvesting energy and materials and converting them into offspring, has evolved senses, structures that are sensitive to some parts of reality and able to transduce that into patterns of nerve signals. The senses that animals have acquired are those that are physically possible, and provide the animal with the useful information that it needs to survive, compete, and reproduce. Animals do not sense parts of reality that are not useful for these pursuits. Gathering information is important, but ignoring irrelevant information is equally so. Thus, each species has a sensory reality that is dictated by its ecology. The realities of species can be represented as sets. The degree to which the sensory sets of two species overlap is governed primarily by their ecology and by their evolutionary relatedness. The smaller the time interval since the evolutionary divergence of two species, and the more similar their ecologies, the more similar their realities will be. For animals there is no single reality.

Seeing with Ears

Geerat Vermeij, an evolutionary biologist at the University of California at Davis, has shown that over evolutionary time, life inexorably becomes more dangerous and competitive, as organisms become more effective at competition and predation, and acquire better defenses. For example, flight in insects likely evolved in response to a world made increasingly dangerous by insect-eating animals. Then a descendant group of the dinosaurs, the birds, evolved flight, and insects became the main menu item for many bird species. In response, some insects escaped bird predation by an evolutionary change in their daily activity schedules, resting during the day and becoming active only at night, when birds, that hunt using vision, could not see them. As a result, a huge potential advantage appeared for any animal group that could evolve to exploit the insect-filled nighttime air. Into this huge vacant niche stepped a group of tiny mammals: the bats.

Unlike birds, mammals were already nocturnal, so the bats had only two remaining changes to make: develop flight, and develop a way to see in the dark. They did both, and became (after rodents) the second most successful mammal group. Most bats fly at night and pluck insects from the air. The fact that there are so many bats (both many species, and, within species, millions of individuals) means that they are accomplished at a seemingly impossible task. In the mid-twentieth century, biologists who wanted to settle the question of whether bats "see" their prey, or whether they simply fly through the insect-filled air with open mouths, took a direct approach: they shot individual bats at known times after they emerged from their roosts in the evening and weighed their stomach contents. The results were conclusive. Within an hour after emerging from its daytime roost, a bat has packed its stomach with a mass of insects equal to about one-fifth to one-third of its body mass. This means that a bat catches about 500 insects per hour, or one insect every seven seconds. Simple calculations showed that these capture rates could not result from a bat simply flying with its mouth open. Rather, the bats are detecting, pursuing and capturing individual insects.

Although bats act as if they are able to see insect prey in the dark, as well as to see obstacles and other features of the environment, they have tiny eyes and weak vision – optometrists would classify most bats as legally blind. Thus, as early as the eighteenth century, biologists postulated that bats might somehow be able to "see" with their ears. However, real proof came only in the mid-twentieth century. The proof resulted almost entirely from the brilliant efforts of one man, Donald Griffin, who combined the observation of bats in nature with new technology that allowed him to detect and to record the sounds, inaudible to humans, which bats produce. Griffin discovered the sensory process known as *echolocation*.

What we call sound is our sensory interpretation of wave energy, produced by the vibration of an object in air, which propagates through the atmosphere at about 343 meters per second. You can produce a visual representation by dripping drops of water into a still pool: waves propagate from the source of the disturbance. Depending on the size of the drops and the height from which you drop them, the waves will have a specific height, or amplitude. Depending on how rapidly you let the drops fall, the waves will have a specific distance, or frequency, from crest to crest. When a sound wave passes by any location, the air pressure there rises slightly and then falls back to atmospheric pressure.

In the terrestrial mammals, the hearing sense operates by detecting the tiny transient fluctuations in atmospheric pressure that are produced by sound waves. Differences in amplitude are perceived as differences in loudness, and differences in frequency are perceived as differences in pitch. The eardrum, a thin membrane that separates the back of the throat from the atmosphere, flutters as sound waves strike it. Resting against the inside of the eardrum is a chain of three little bones: the malleus, incus, and stapes. Vibration of the eardrum is passed along the chain of bones to a membrane on the cochlea, the tiny conch shell-shaped

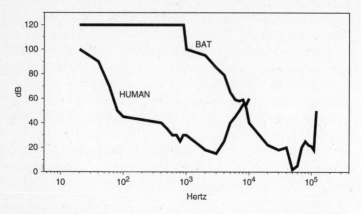

Figure 2.1 Approximate auditory tuning curves for humans and for an echolocating bat species

organ of hearing. The cochlea is where patterns of mechanical vibration are transduced into patterns of activity in the auditory nerve, which carries the information to the brain. The cochlea is the principal bit of anatomy where the auditory sensitivity characteristics of a species are created. For any species, hearing physiologists can construct a tuning curve, where sound frequency is represented on the *x* axis and the minimum sound pressure at which a response from the cochlea can be detected is represented on the *y* axis. A tuning curve shows both the range of frequencies that a species can hear and the range of frequencies that a species hears best. Tuning curves usually have the shape of a shallow 'U'. For humans, the tuning curve shows the ability to hear sounds from 20 to 20,000 Hz (Hz = Hertz, or number of wave crests per second) with the bottom of the U, the most sensitive hearing, centered on 2,000 to 4,000 Hz, a region where most of the energy in human speech sounds occurs. For bats, the curve shows the ability to hear from 20 to 120,000 Hz, with the greatest sensitivity at 40,000 to 80,000 Hz.

Bats are able to hear very high frequency sounds because they produce such sounds to echolocate. A bat that has emerged from its roost, flying at perhaps 20 meters above the ground, emits a series of clicks. Like all mammal voices, the clicks are produced in the larynx. Each click is of short duration, about 0.0012 seconds; is very loud, about 110 decibels, a sound pressure equivalent to that produced by a pneumatic drill breaking up pavement; is very high-pitched, at about 40,000 to 100,000 Hz; and typically is a glissando (a continuous slide between two notes), with a steep frequency drop of 30,000 Hz or more. Griffin referred to these short, loud, high-pitched, frequency-modulated screams as "cruising pulses." A cruising bat emits these at a rate of about eight per second. While the bat is screaming, it protects its own hearing by contracting muscles that pull sideways on the malleus and the stapes, partially unhinging the chain of ear bones. The bat then turns its ears back on, re-hinging its ear bones, and listens for the echo of its scream, using the information present in the echo to produce a picture of its immediate surroundings.

When a cruising bat detects an object of interest (perhaps a moth, perhaps a spitball shot into the air by Donald Griffin with a glass tube blowgun), it flies in that direction, and begins to increase its click rate and to shorten the duration of each click, building pictures of the immediate environment at an accelerating rate. As the bat closes on the target, it emits what Griffin called a "buzz," in which each click lasts just one millisecond and clicks are emitted at 200 per second.

Biologists continue to make discoveries about the capability of the bat echolocation sense. They have had some success in working out the general algorithms by which echo processing occurs, and have mapped the brain areas that receive and process information from the auditory nerves. We now know enough to evaluate the eighteenth-century conjecture that bats must "see" with their ears. The conjecture was correct. Bats do indeed have a sense that works in the dark and that rivals the detail provided by vision.

What would this sense feel like if we possessed it? We can only guess, like a congenitally blind person trying to imagine what vision feels like. Bat echolocation illustrates the main point of this chapter. Each animal species has its own sensory tuning and sensory processing abilities and hence its own reality.

Moth Ears

Many of the nocturnal moths that bats pursue exemplify the principle of inattention to irrelevant information. These moths have evolved an ear which consists of an eardrum on each side of the body below the wings, and exactly two nerve cells per side that report to the moth's brain. This simple ear has a flat tuning curve across a wide frequency range from 20,000 to 100,000 Hz – the range of bat echolocation clicks. The moth does not, in effect, care which sound frequencies are present, and it does not hear sounds below 20,000 Hz. The information relevant to the moth is whether a bat is nearby and, if so, how close. The simple ears allow a moth to detect a bat at a range that is greater than the range at which the bat can detect the moth, and the two nerve cells per ear also provide adequate information on the location of the bat. A moth that makes such an early detection (the nerve cells of the ear facing toward the bat will be active and the nerve cells of the facing-away ear will not) turns abruptly and flies directly away from the bat. If the moth is detected, its ears report that the echolocation clicks are louder and more frequent. Beyond a certain threshold, the moth's response abruptly changes; it drops to the ground, in a chaotic way that makes its flight path unpredictable. Although they are adversaries locked in an escalating predator–prey struggle, bats and moths could not have more dissimilar auditory realities. For bats, auditory processing provides a rich, vision-like picture of the immediate surroundings in the

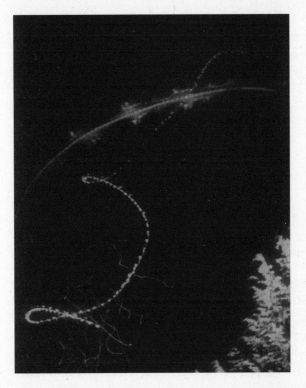

Figure 2.2 An instant in the summertime night sky, captured by Ken Roeder. The fighter jet-like arc indicates the path of a foraging bat. The looping chain indicates the path of a moth that detected the close approach of the bat and initiated its chaotic descending escape. The faint traces are those of other insects.

nighttime air; for moths, auditory processing is like the check oil pressure light in an automobile dashboard.

Fly Eyes

Eyes are sensory structures that are adapted to transduce light energy into useful patterns of nerve activity that the brain can use to build an accurate representation of its environment. Light is electromagnetic radiation, which has both particle and wave characteristics. The wave characteristics allow us to characterize types of electromagnetic energy by wavelengths. What humans perceive as visible light has wavelengths from 400 to 700 nanometers (a nanometer (nm) is one-billionth of a meter). At wavelengths shorter than 400 nm are the forms of radiation called ultraviolet, X-rays, and gamma rays, with wavelengths down to 0.01 nm. At wavelengths longer than 700 nm are the forms called infrared, microwave, and radio wave, with wavelengths up to 1,000 meters or more. Animal eyes have evolved sensitivity to electromagnetic radiation in the narrow band that humans call visible light because this range represents the majority of solar radiation that reaches the Earth, and because radiation in this range can provide useful information about objects in the environment. Like sound waves, electromagnetic radiation in this range bounces off objects and thus provides information about what is out there. Because light travels much more quickly than sound, and because light has such small wavelengths, it represents an incredibly rich source of information. More than any other sense, light provides highly detailed, accurate information about the presence of objects, their motion, and their positions as an animal moves near them. A peregrine falcon may reach speeds close to ninety meters per second as it dives at its prey, but light travels at about three hundred million meters per second, so the falcon's eyes give completely accurate reports about the distance between it and its prey.

All animal vision basically operates in the same way. Light-sensitive cells in the eye contain visual pigments, molecules that undergo a structural change when they absorb a photon, the particle unit of light. The shape change starts a chemical cascade that results in the production of a nerve signal. Among animals, variety in the specific wavelength sensitivity of visual pigments and in the optical equipment that delivers light to the pigments produces variety in visual reality.

All vertebrate animals (animals with a vertebral column, or backbone: fish, amphibians, reptiles, birds, mammals) have the same basic visual equipment. They have two spherical camera-type eyes, with a pupil (a hole) to admit light, a lens to focus images on the photosensitive retina, muscles that deform the lens to focus at different distances, and six muscles that rotate the eye in its socket. The ability to discriminate between different wavelengths of light – to perceive what humans call color – depends on the variety of visual pigments in the cells of the retina. Most of the vertebrates have more kinds of visual pigments than we do, and thus they see more colors, including colors into the ultraviolet.

The other major eye design of animals exists in the arthropods: the spiders, scorpions, centipedes, millipedes, lobsters, crabs, shrimp, and insects. Most arthropods see with what is called a compound eye. A compound eye, such as the eye of that housefly that you crushed at the beginning of this book, is composed of several thousand individual units called *ommatidia*. Each ommatidium is a long, slender tube with an exterior lens and interior cells that contain light-sensitive pigments. Imagine a closely packed bundle of drinking straws with one end of each stuck into a baseball-sized lump of clay. Each straw will point in a specific direction. Because the compound eyes of arthropods often make up more than one-half the surface area of the head, an individual animal has ommatidia that point in almost all directions. Thus, even though your hand approached the fly from behind, she detected you easily.

The compound eye is more democratic than the vertebrate camera eye, with its dominant *fovea*, the area where receptor cells are packed most closely together and where interpretive representation (nerve cells) in the brain is most rich. In the vertebrate eye, activity in most of the retina away from the fovea is used to tell the brain whether there is something worth pointing the fovea at. The vertebrate eye throws away a lot of information to gain the very high image resolution that the fovea creates. In contrast, the compound eye gives equal representation to the output from

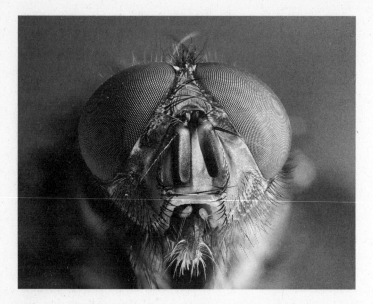

Figure 2.3 Scanning electron micrograph of the head of a fly. In the enormous compound eyes, each dot represents an ommatidium.

each ommatidium. The resulting view of the world is thus something like a pointillist wide-screen IMAX. The compound eye has weak resolution but it is superbly good at detecting motion. Florescent lights flash 120 times per second under the rapidly flip-flopping polarity of alternating current. A housefly sees the separate flashes. She is so good at detecting motion that you can only get at her with the raw speed provided by your fly swatter.

Tactile Foveation

Camera eyes solve the problem of processing abundant information by assigning most of the retina the menial task of reporting whether there is something to look at, while assigning a small area of the retina, the fovea, the job of analyzing the detail of images. In the fovea, light receptors are packed together at high density. A correspondingly large area of the visual brain is dedicated to processing signals from the fovea. Thus, camera eyes almost constantly make small twitching adjustments to point the fovea at specific spots in the field of view. These movements are called *saccades*. We do not notice other people's saccades because the movements are small and quick: the typical duration is three to six hundredths of a second.

A similar kind of sensory processing, the devotion of one part of a sensory apparatus to detailed analysis and the use of the remainder of the apparatus to identify analysis targets, occurs in a burrowing mammal, the star-nosed mole. Like other moles, this species is adapted to find and eat subterranean invertebrates. Moles use their powerful forelimbs to swim through soil, constructing a burrow system, which they patrol for food. Within this narrow, specialized niche, star-nosed moles have become even more specialized; their diet consists of food items that are underground and also tiny.

Eating tiny food items is effective only if the time spent collecting and eating each one is brief. What counts in the economy of nature is the rate of energy acquisition, the calories consumed per second of foraging. A species can become a specialist at eating tiny food items, but only if the handling time per item is brief. A blue whale, the largest living animal on earth, eats tiny food items. The whale achieves a small handling time by taking a volume of sea water the size of a school bus into its mouth, and then pushing the water through its baleen (filtering) plate, leaving behind a few million planktonic plants and animals to swallow.

A mass filtering strategy is not available to an animal that swims through soil. Star-nosed moles find and consume prey items one at a time, so if they are to specialize in tiny food items, the handling time for each item must be extremely brief.

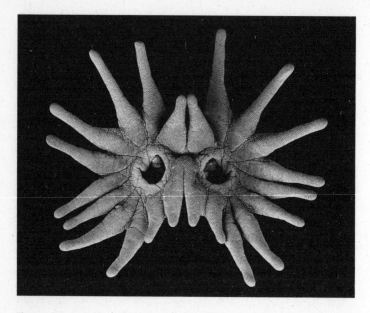

Figure 2.4 Scanning electron micrograph of the star of a star-nosed mole. Each individual bump represents a touch organ. Photograph courtesy of Ken Catania.

The moles achieve really rapid handling times by touch, using the unique stars on their noses. A star-nosed mole has a slightly upturned snout. Surrounding each nostril is a crown of fleshy tentacles, which are studded with thousands of touch receptors. The bottom tentacle, number eleven, which rests just above the mouth, is the touch equivalent of a fovea. The area of the brain that processes the signals from this tentacle is much larger than the areas devoted to the other tentacles.

When the outer tentacles of the star touch a potential food item, they move to bring that item to tentacle eleven in an extremely rapid movement, as quick as or quicker than a saccadic eye movement. This rapid evaluation creates an average handing time of about two-tenths of a second, making it possible for star-nosed moles to be small food item specialists. Over half of the star-nosed mole's somatosensory cortex, the brain area that is devoted to the analysis of nerve reports from touch receptors, is occupied by areas that process information from the star.

Electrolocation

We can go further into the realm of the unimaginable by considering the ability of some animals to detect electric fields. Sharks and rays use this sense to find prey and possibly to navigate. On the snout, around the mouth, and above and below the eyes of these fish is an array of clearly visible dark pits. The short pits, open to the surrounding water, are filled with a jelly-like material and have receptor cells at the bottom. These are the *ampullae of Lorenzini*, first described by Stefano Lorenzini in 1678, but not proven conclusively to be electroreceptors until 1971. The receptors detect voltage differences between the pore opening and the base, and are extraordinarily sensitive. They are able to detect a difference of one hundred millionth of a volt at one meter. Because the heart, breathing muscles, and skeletal muscles of animals produce small electric fields, a shark can detect a flounder, motionless and buried beneath the sand, at a distance of 30 centimeters. The sensitivity of the ampullae of Lorenzini is so great that a shark can detect the direction of the earth's magnetic field and may possibly use this information to guide long-distance trips.

In the murky rivers of the Amazon basin and central Africa, two distantly related groups of fish, the *Gymnotiformes* (South America) and the *Mormyridae* (Africa), have taken electroreception to another level. These animals perform the electrical equivalent of bat echolocation. These so-called "weakly electric" fish use specialized blocks of muscle to produce an electric field around the body; they then detect disruptions in the field. When it is near to an object, the fish experiences a kind of electric shadow that falls on the body, and it uses this information to build a picture of the immediate environment. The weakly electric fish also communicate with each other using pulsed electric discharges.

The Sensory World of a Tick

A final example of multiple realities is revealed by the sensory world of a female tick. A female tick needs a meal of mammal blood to provide the nutrition needed for her eggs to mature. A female, having found an elevated perch on a blade of grass, becomes insensitive to everything in her environment except a single chemical compound. She may wait, motionless, for a decade or more, ignorant of the rising and setting sun, breezes, coyotes howling, birds singing, rain and snow, and the apparent passage of time. All her senses, except her chemical sense, are turned off, and her brain is waiting for the single signal that indicates the detection of a particular compound: butyric acid. This compound is one of the salient constituents of unwashed armpit odor, and is produced in the glands at the base of the hair follicles of all terrestrial mammals. Butyric acid is an infallible indicator that a mammal is close by. When the female tick perceives this compound, she jumps. If she is fortunate, she lands on the mammal. Then she loses her sensitivity to butyric acid and gains sensitivity to heat. By moving up a gradient toward hotter temperatures, she reaches the skin, where she will find her blood meal. For a tick, reality is contracted to the perception of three stimuli: butyric acid, which releases her jump; light, which directs her climb; and heat, which guides her to the skin.

At the beginning of the twentieth century, a German biologist, Jakob von Uexküll, used the example of a female tick to support his thesis that each animal species, because of its particularly designed and tuned sensory system, has its own sensory world, its own *Umwelt*. What we have learned about the sensory worlds in the century since von Uexküll has verified this fundamental insight. Each species really does have its own Umwelt, its own perception of reality, which provides individuals with the information that they need to survive, compete, and reproduce. Just as important, the Umwelt excludes information that is irrelevant to these goals. There is a single reality, but no animal species perceives all of it. All animal species, taken together, do not perceive all of reality. Some parts of reality are just not useful to an organism that is trying to harvest energy efficiently and maximize the number of offspring that it can produce.

3
Adjusting Priorities

The great French naturalist Jean-Henri Fabre (1823–1915), a pioneer in the study of insect behavior, described many examples of the apparent wisdom but stupefying rigidity of behavior that is produced by pencil point-sized brains. Fabre made detailed observations of nesting behavior in several digger wasp species. When she is ready to lay an egg, a female digger wasp excavates a tubular burrow in soil. She then hunts for prey, such as a cricket. She stings the cricket, inducing paralysis. She then drags the living cricket head first into her burrow, and lays an egg at the base of one of the cricket's legs. She exits the burrow, scrapes sand and pebbles into the burrow mouth to seal it, and flies away, never to return. Underground, the living cricket provides the wasp larva with abundant fresh meat.

Fabre performed an experiment with a wasp in his home laboratory. While the female was kicking sand to fill a burrow that she had just provisioned, Fabre pushed her aside and used a knife and forceps to clear the partially-formed burrow plug. He then removed the cricket, with the single wasp egg glued to the base of its leg, and put it in a box. The female wasp stood nearby while Fabre did this. When Fabre took his hands and instruments away, the female entered the burrow, emerged soon after, and resumed the plugging operation. When the plug was complete, she flew away. Although the female obviously had the opportunity to perceive that the cricket and her egg were no longer in the burrow, she continued with the behavioral sequence at the point when Fabre had interrupted it. In this and hundreds of other experiments, Fabre illustrated how insects, with their tiny brains, can exhibit complex, goal directed behavior through chains of instinctive responses. The stimulus that indicates the completion of one task in the chain initiates the next set of responses. Once a female has switched into the next sequence, she can't stop, and she can't go back, as Fabre illustrated.

Rigid, instinctive responses are also found in animals with relatively larger brains. The phenomenon of brood parasitism in birds provides an example. Brood parasites are bird species that rely on other bird species to rear their young. A brood parasitic female, such as a cuckoo or a cowbird, searches for the nest of an appropriate host species. The parasitic female lays an egg in the host nest and departs. The parasitic egg hatches first, and in some species such as cuckoos, the newly hatched chick backs up to each egg, maneuvers it onto its shoulders, and pushes it out of the nest. Now there is a single parasitic chick in the nest, and it demands food by cheeping and throwing its beak wide open, as the hosts' chicks would have done. The host parents respond to this irresistible stimulus by feeding the parasitic chick. Even when the parasitic chick has grown to be larger than the host parents, the parents work valiantly to keep up with its food demands, standing on tiptoe to stuff food into its open beak. Nest parasitism has evolved many times because of the predictable rigidity of a bird parent's response to a begging chick. The host parents are trapped, condemned by their unvarying response to a stimulus to a reproductive season wasted in the service of another species.

Animals may be stimulus–response machines in some circumstances, but in many others they are capable of adjusting priorities so that the response to a stimulus varies. The ability to adjust responses to stimuli should evolve whenever the environment provides reliable contingencies that provide a reproductive advantage to the individual that can adjust. There are many illustrative examples of this, which is the province of the sub-discipline of animal behavior known as *behavioral ecology*.

How Fat Should a Great Tit Be?

A few species of seed-eating birds, such as chickadees and great tits, have evolved the ability to stay at home during winter. These birds avoid the costs and risks of flying a thousand miles or more twice each year, but they incur the costs and risks of surviving through winter. Every winter night, small birds risk dying. During the day the bird must find and consume enough food to accumulate a fat reserve. At night, as the bird huddles, motionless, its feathers fluffed out to provide as much insulation as possible against the freezing winter, it uses the stored fat as the energy source to stay warm and alive. Even though fat is compact, it weighs something, and weight is always a critical concern for a bird. Every gram of fat that a bird stores is an additional gram it must carry when it walks or flies. A heavier bird is also a slower, less agile bird; one more vulnerable to predators. Accumulating an extra gram of fat also requires that a bird is out and exposed to predators while it searches for food.

The term that behavioral ecologists use to depict such a situation is "trade-off." A great tit faces a trade-off between starvation and predation risk. In the laboratory, where researchers can control the birds' access to food using electronically controlled doors on feeders, putting the birds on an unpredictable feeding schedule causes them to eat more and to gain weight. They perceive a starvation risk, and so add fat. Putting the same, heavy birds back on a predictable feeding schedule causes them to eat less and to lose weight. If there is no starvation risk, it is better to shed the extra weight, and lower the risk of being caught by a predator.

The Varying Value of a Male Child

In the science fiction thriller *Alien* and its sequels, humans meet arthropod-like creatures that insert an egg into a human. The egg develops inside the human and eventually a juvenile alien emerges, bursting through the human's abdominal wall. The aliens resemble arthropods because the original writers, Dan O'Bannon and Ronald Shusett, took the idea for a species that grows inside another species from real arthropods: the parasitoid wasps.

A parasitoid wasp lays her eggs on or inside the body of another insect larva or pupa, such as the caterpillar of a moth or butterfly or the maggot of a fly. The eggs hatch and the wasp larvae feed on the living body until they have grown enough to pupate, emerging from their now consumed and dead host as new adults. There are thousands of parasitoid wasp species, each specialized for a narrow range of hosts. In agriculture, many parasitoid species are known as valuable controllers of crop pests.

Typically, the male wasps of a brood emerge first. The males have stunted wings and are incapable of flight. A male wasp waits for the emergence of the females, mates with one or more of them, and then dies. The newly mated females then fly away to search for a host.

Because of this very specific life cycle, a female wasp that has found a fresh larva faces an unusual problem. If she is the only female to lay eggs on the host, all her sons will compete with each other to mate. This is wasteful. The female could get more reproduction out of her brood if she could somehow produce mostly females, with just a few males to perform all the matings. In the wasps this circumstance is probably what led to the evolution of a specialized mechanism of sex determination known as *haplodiploidy*.

When she mates, a female wasp stores the sperm in a compartment of her reproductive tract called the *spermatheca*. The sperm stored there can live indefinitely. As an egg passes down the oviduct, the female releases sperm from the spermatheca to fertilize the egg. A fertilized egg is diploid (having two copies of each chromosome, one from the father, and one from the mother) and will develop into a female. If the mother withholds sperm as the egg passes the spermatheca, the egg remains haploid (one copy of each chromosome from the mother only), but still develops, and becomes a male. Haploid = male; diploid = female: haplodiploidy. Through this method of sex determination, a parasitoid wasp female can control the sex ratio of her brood.

In nature, hosts may be scarce, or they may be concentrated in space (for example, fly maggots in a bird's nest). Thus, there is a risk that more than one female wasp will lay eggs on a host. As the number of separate broods developing in a host increases, the fitness value to the mother of a male offspring increases. A male offspring can also mate with emerging females from other broods, and males from other broods increase the level of competition. In 1967, the famous theoretical biologist William D. Hamilton referred to this situation as *local competition for mates*, and predicted that the sex ratio of the brood that a parasitoid wasp female lays should vary with the number of other females that have already placed broods in the host. His prediction was that when a female had the host larva to herself, she would produce about 20% sons, and that as the number of other broods in the host increases, the sex ratio that the female produces would converge on 50% sons.

Hamilton's prediction has been confirmed, in a number of species, in nature and in the laboratory, where the conditions that a female experiences when she lays her brood can be controlled. Females do lay more male eggs as the level of local competition for mates that her brood will face rises. How do females "know" to lay more male eggs? In other words, what are the stimuli that influence the sex ratio that a female produces? In one species, *Nasonia vitripennis*, a tiny wasp that lays eggs on fly pupae, we know that the presence of other females has an effect. If other females are nearby when a female lays her brood, she produces more males. Even more important is the presence of other eggs. *Nasonia* females lay eggs on fly pupae. The pupa, a larval fly in the process of metamorphosing into an adult, is surrounded by a puparium, a tough shell. During a walking tour of a pupa after she finds it, a female wasp may detect the holes that other females have drilled through the puparium. After she has drilled her own hole and inserted her long ovipositor to deposit eggs on the body of the pupa, she may be able to touch or smell other eggs. If she detects the eggs laid by other females, then she becomes less likely, as an egg passes toward her vagina, to contract the muscles that control the spermatheca, and allow the egg to be fertilized. She lays more unfertilized, haploid eggs that will develop into males.

The Infanticide Clock

Natural selection structures the brains of species so that individuals act as if they are constantly asking themselves the question, "what should I be doing at this moment to maximize my lifetime reproductive success?" It is now necessary to make the first of two adjustments to improve the accuracy of that statement. In the quotation, substitute "fitness" for "lifetime reproductive success." What is the difference? Lifetime reproductive success is simply the number of offspring that an individual produces in its lifetime. Fitness is that number divided by the lifetime reproductive success of the most successful class of individuals in the population. In other words, fitness measures how much an individual reproduces compared to all other individuals in the population. Biologists want to measure fitness because it predicts the direction and rate of evolutionary change of traits.

Because fitness is relative, an individual can increase its own fitness in two ways. First, it can reproduce more. Second, it can diminish the reproduction of others. Thus, individuals often act in ways that, from the viewpoint of human societal norms, are brutally selfish and repugnant. An example is infanticide: the killing of infants by adult males. Infanticide is well documented in many animal species and usually occurs in the same context, when a male that is not the sire of an infant comes into close, often prolonged, physical proximity to the mother and the infant.

Infanticide by males tends to be found most often in species that have a social structure in which a single adult male has exclusive access to a group of females and this male is periodically deposed. However, male infanticide also occurs in species with more fluid social structures. It occurs in many rodents, including house mice. The house mouse has accompanied humans around the globe, and it is found in terrestrial habitats worldwide, including remote oceanic islands. It is a superb colonist, but it also can live at locally high densities. The species shows many finely tuned reproductive adaptations, including a fine-tuned control of infanticide.

In the laboratory, we can study the cues that influence infanticide by male mice. If researchers place a pink, hairless, blind, day-old mouse pup in a cage with an adult male mouse, they observe one of two reactions. The male may approach the pup, rattle its tail aggressively, and maul and bite the pup until it is dead. Or, the male may approach the pup, lick it, and huddle over it protectively. In most instances, we get the Mr. Hyde mouse. But, if researchers have recently provided the male with a female and he has copulated, we probably get the Dr. Jekyll mouse. If the male copulates and the researchers test him on successive days, a clear pattern emerges. From zero to seventeen days after copulation, the response to a pup is infanticide. From eighteen to twenty days, the male's response to a pup abruptly changes to parental and protective. Gestation in the house mouse is eighteen days long. Thus, copulation by a male starts a timer that goes off when a litter sired by that male could be born. When the timer goes off, infanticidal reactions to pups are inhibited for about a month.

Hot and Cold Lizards

Agama savignyi is a small, day-active lizard species that lives mainly on the desert sand dunes of the Middle East and adjacent parts of Africa. Because the species is relatively small, it has quite a few reptile, bird, and mammal predators, and because it inhabits sand dunes, it has limited shelter.

Researchers brought *Agama* lizards into the laboratory to measure the effects of body temperature on running speed. They built a short racetrack, with a rubberized floor for good traction. They chased the lizard, by reaching toward it with a hand, to make it run down the racetrack. Using this technique, they measured the maximum sprint speed a lizard could reach at different body temperatures. Most reptile species have a preferred body temperature, which they attempt to maintain by basking or by seeking shade. At that preferred temperature, an individual's ability to move is best; at higher and lower temperatures, ability to move is impaired.

When the lizards were cold, at a body temperature of about 20°C, they could run at a top speed of one meter per second. Running speed increased with increasing body temperature, up to the preferred body temperature of 35°C, at which the lizards ran at three meters per second. The researchers also discovered that the responses of lizards to the threatening hand varied with temperature. Warm lizards ran away almost immediately, but cold lizards did not. Although cold lizards could run slowly, their principal response to the researcher's hand was to adopt a threatening posture, to gape, exposing the teeth, and to lunge and bite. The researchers reported that these bites were painful. *Agama savignyi*, a sand dune-living lizard with limited access to shelter, modulates its responses to a threat based on its temperature and consequent ability to run well.

Peeking Ducks

Biologists do not entirely know what sleep is for, but we do know that most vertebrates and some invertebrates need to sleep. It is also clear that sleep poses a risk for any animal that could be attacked. In some birds, including mallards, the world's most widespread duck species, individuals lessen the risk by occasionally opening an eye as they sleep. This brief eye opening, which David Lendrem called "peeking," does not interrupt the characteristic brain wave activity of sleep, but it allows an animal to rouse from sleep if a threat is detected.

From a bridge over the Thames at Oxford, UK, Lendrem watched mallards as they slept on a wooden jetty that projected into the river. He watched individual birds for two-minute periods, counted the number of peeks, and noted how far out on the jetty the bird was. He found that a duck's distance from shore influenced the rate of peeks per minute. The closer a duck was to shore, and hence the greater its risk of attack by a ground predator such as a cat, the higher was its peeking rate.

A great tit regulates how much it eats depending on its perceived risks of death by starvation and death by predation. A parasitoid wasp female alters the proportion of male eggs that she lays on a host depending on her perception of the number of competing broods that might already be there. A male house mouse has a brain clock that tells him when it is advantageous to be nice to a mouse pup. Desert lizards change their responses to a threat depending on their body temperature. Mallards scan the environment more if they are asleep in riskier locations. These are just a few examples of a general finding confirmed by the study of behavioral ecology in the past twenty years. Animals can evolve the ability to adjust priorities, to modify their responses to a stimulus depending on context, when there is a fitness advantage to be gained by doing so. Biologists write mathematical models to predict optimal switching (for example, Hamilton predicting the optimal brood sex ratio under differing levels of local competition for mates), and often, they find that animals in nature adjust their behavior, with impressive precision, and stay close to the predictions.

4

Brains and Glands

Eating improperly prepared fugu will kill you. You will remain fully conscious as you become paralyzed and your breathing stops. *Fugu* refers to several species, commonly known as blow-fish or pufferfish, which are used in Japanese cuisine. These fish are slow swimmers, propelling themselves by movements of their fins. They compensate for their vulnerability in two ways. First, they are able to inflate. They can quickly fill their stomach with enough water to swell into a spherical shape. This, combined with a spiny skin, makes them difficult or impossible to swallow. Second, these fish store, in some organs and in the skin, a poison produced by bacteria that live beneficially within them. This poison, tetrodotoxin or TTX (the fish belong to the order *Tetraodontiformes*) is gram for gram, about a hundred times more lethal than cyanide.

Tetrodotoxin induces paralysis because it binds to and inactivates the sodium channels in nerve and muscle cell membranes. These channels, which regulate the movement of sodium across the cell membrane, are the basis for signaling in the nervous system and for the initiation of muscle contraction. The action of sodium channels, in concert with other ion channels and pumps, creates the two fundamental states of a nerve or muscle cell membrane: resting potential and action potential.

Potentials

A nerve cell, like any other cell, has an outer cell membrane, which is a double layer of fat. The double fat layer prevents atoms or molecules from simply passing into or out of the cell; it is the primary basis for the cell's ability to regulate what is crossing the membrane. Embedded in the cell membrane are channels, proteins that are specialized to act as pores for specific atoms or molecules. Also embedded in the membrane are proteins that act as pumps, to move atoms across the cell membrane in one direction. A nerve cell has a cell body, from which project a number of short branches, *dendrites*, and a single long branch, the *axon*. Dendrites receive signals and the axon transmits signals.

To make electrical recordings from a single axon, researchers set up electrodes (insulated wires with bare tips) that are connected through a voltmeter. This was first done in squid, which have a giant axon, 0.5–1.0 mm in diameter, which drives the contraction of the squid's mantle when a fast escape is needed. Imagine a squid giant axon set up in a dish of seawater, in which lie two electrodes, attached to a voltmeter. If you leave one electrode in the water and touch the other to the surface of the axon, the voltmeter reads zero. If you push the touching electrode through the cell membrane, into the interior of the axon, at the instant of penetration, the voltmeter registers a change of about −70 mV (mV = millivolt, or one-thousandth of a volt). The inside of the axon has a negative charge with respect to the outside. If the electrode is left undisturbed, the voltmeter continues to read a steady −70 mV. This charge difference between the inside and the outside of the cell is the cell's *resting potential*.

Resting potential arises primarily through the electrochemical equilibrium of potassium ions ($K+$). Electrochemical equilibrium means that there are two effects, one electrical, and one chemical, which control the local concentration of a charged ion. The electrical effect is the attraction of opposite charges: positively charged ions move toward negative charges, and vice versa. The chemical effect is diffusion. Ions such as sodium ($Na+$) or potassium in a solution of water (as they are inside and outside the cell) move until they are evenly distributed. If you drop a teaspoon of table salt (sodium chloride) into a glass of water, the sodium and the chloride ions soon will be evenly distributed; no part of the water will be saltier than another. Ions in solution move from an area of high concentration to areas of low concentration.

Inside the axon, a permanent source of negative charge results from protein molecules that have negative charges. Embedded in the cell membrane are big proteins that act as channels and pumps. There are Na+ channels, K+ channels, and a sodium–potassium pump. The pump works constantly, using stored energy in the cell to move Na+ out of the axon and K+ in. When the axon membrane is at resting potential, the Na+ channels are closed: no Na+ ions can get through. Some of the K+ channels are open. As a result of the pump and the channel settings, the concentration of Na+ outside the axon is much higher than the concentration inside. The Na+ outside cannot diffuse to the area of lower concentration inside, because the Na+ channels are closed. The concentration of K+ is higher inside than outside. Because some K+ channels are open, K+ can diffuse out of the axon. As each K+ ion leaves, the balance between positive and negative charge inside the axon becomes more negative. Eventually, equilibrium occurs at the point when, for each K+ ion that diffuses out, moving down a concentration gradient, another K+ ion moves in, pulled by the negative charge. This electrochemical equilibrium results in a small negative change of about −70 mV inside the cell. This is called the resting potential, but the cell is definitely not resting. It is constantly running its sodium–potassium pump. The negative charge inside the axon is called the resting potential because, unless it is disturbed, the axon membrane keeps this steady charge difference between inside and outside.

How does an axon transmit a nerve signal, an *action potential*? The key to this event is the nature of the Na+ channels. These channels are sensitive to the charge inside the cell. At the resting potential, the channels are closed. When the cell becomes somewhat depolarized, moving from the resting potential of −70 mV inside to about −50 mV inside, the Na+ channels begin to open. Due to the sodium–potassium pump, the concentration of Na+ outside the axon is much higher than the concentration inside. When the Na+ channels start to open, Na+ diffuses into the axon. As Na+ moves in, the negative charge inside the cell is erased, and this causes the Na+ channels to open further. The Na+ channels are described as "voltage-gated." Because the Na+ channels act this way, once the membrane reaches the point at which inward movement of Na+ starts, there is no going back. The point at which Na+ starts to move inward, at a membrane charge of about −50 mV inside, is called *threshold depolarization*. Once threshold depolarization is reached, the accelerating inrush of Na+ is inevitable.

At any place along the axon where threshold depolarization occurred, the axon has a positive charge (+50 to +100 mV) inside, because of all the Na+ ions that have rushed in. Immediately adjacent, the axon membrane is at resting potential: about −70 mV inside. Such opposite charges so close to each other cause a flow of electric current sufficient to induce threshold depolarization, the inrush of Na+, and the reversal of the membrane's polarity at the adjacent spot. This causes threshold depolarization at the next adjacent spot, and so on. This mechanism, like falling dominoes, is how action potentials are propagated along an axon.

At the instant at which the inward flow of Na+ made the axon positively charged inside, the Na+ channels begin to close, due to a complex mechanism in the big protein molecule that is the channel. Also, K+ is no longer held inside by negative charge, so K+ diffuses out, restoring the resting potential. The whole event − the inrush of Na+ that reverses membrane polarity and the outrush of K+ that restores resting potential − occurs in about one-thousandth of a second. An action potential is a very brief flip-flopping of membrane polarity that propagates along the length of the axon, at speeds of up to 120 meters per second.

This ending point is called a *synapse*. A synapse comprises a pre-synaptic membrane, where the axon ends, a synaptic cleft (the tiny gap), and a postsynaptic membrane on the receiving cell. When an action potential reaches the synapse, it triggers the opening of calcium (Ca++) channels in the presynaptic membrane. The inward movement of Ca++ sets off a chain of events that end when synaptic vesicles, small spherical storage bags inside the cell, move to the presynaptic membrane, fuse with it, and discharge their contents into the synaptic cleft. The compound that is released into the synaptic cleft is one of a class of compounds called neurotransmitters. For example, at the neuromuscular synapse, the neurotransmitter is acetylcholine. Acetylcholine diffuses across the synaptic cleft and binds to receptor proteins embedded in the postsynaptic membrane. Binding induces the opening of Na+ channels and the initiation of an action potential that sweeps across the muscle cell membrane and then along membrane conduits, deep into the fiber. In response to the action potential, the machinery of fibril shortening starts, and the muscle cell contracts.

A one-to-one relationship between an action potential arriving at the synapse and the muscle cell response is maintained by digestive proteins in the synaptic cleft that rapidly break down the bound acetylcholine. If the muscle cell is to continue to contract, it must receive more action potentials. The digested components are taken up by the presynaptic membrane, re-synthesized to acetylcholine, and repackaged into vesicles. The enzyme that breaks down acetylcholine is called cholinesterase. Some of the most potent insecticides and human weapons of mass destruction (such as the nerve toxin, VX) are compounds that block cholinesterase.

Motor units vary in size. In small motor units, the motor neuron innervates just a few muscle cells. In large motor units, the neuron innervates a few hundred. When the muscle fibers of a small motor unit contract, there may be no visible movement, even though there is a change in tension across a joint. When animals move, they typically activate small motor units first, and recruit larger motor units only if needed. In addition, the force that muscle cells produce depends on the patterns of stimulation that they receive. If action potentials arrive close together, the muscle cell smoothly increases the amount of force that it produces until it reaches a plateau of maximum force: *tetanus*. When motor neurons are active, they typically deliver sets of action potentials that are packed in time sufficiently to make the muscle cell advance toward tetanus, but with durations shorter than what would be needed to create it.

Thus, very fine control of movement is available to the nervous system. This is evident when we observe animals move gracefully, witness an accomplished musical performance, read the facial expressions of other humans, or listen to a song sparrow. Even the clumsiest human, when walking, is immensely graceful when compared to the most sophisticated robot.

Most of the neurons in the nervous system are not motor neurons. They are involved either in reporting information from the sense organs (sensory neurons), or are involved in information processing and decision-making (interneurons). The axons of sensory neurons or interneurons end on the cell body or dendrite of another neuron. The point of ending is a synapse, much like the neuromuscular junction, and the effect of synaptic activity on the target cell is to make it either more or less likely to fire an action potential. The effect that makes a target cell more likely to fire an action potential is called *excitation*. The effect that makes the target cell less likely to fire an action potential is called *inhibition*. Each synapse involves the release of one kind of neurotransmitter (there are dozens of known neurotransmitters), and is either excitatory or inhibitory. At an excitatory synapse, binding of the neurotransmitter to the postsynaptic membrane causes $Na+$ channels to open and that region of the target cell membrane to become depolarized. This is called an excitatory postsynaptic potential (EPSP). As in the neuromuscular synapse, enzymes in the synaptic cleft quickly digest bound neurotransmitter molecules to maintain a one-to-one relationship between input and effect. At an inhibitory synapse, binding of the neurotransmitter to the postsynaptic membrane causes chloride (Cl-) channels to open, making the interior of the target cell at this point hyperpolarized, or more negative with respect to the outside. This is called an inhibitory postsynaptic potential (IPSP). Again, enzymes in the synaptic cleft quickly digest bound inhibitory neurotransmitter.

A neuron in the brain typically has hundreds to thousands of synapses on its cell body and dendrites. Some of the synapses are excitatory and some are inhibitory. An interneuron, then, is a tiny computer. It sums the EPSPs and IPSPs that spread across the cell membrane from moment to moment and, when average membrane polarity at the axon hillock (the synapse-free part of the cell body, where it merges into the axon) reaches threshold depolarization, the cell fires an action potential. If excitation and inhibition are about equal, or if inhibition is predominant, the cell is silent, even though activity at its synapses may be prodigious. A neuron is a simple machine, making yes/no decisions from moment to moment, based on the temporal and spatial sum of excitatory and inhibitory input. The distant ultimate goal of neuroscience is to explain how millions to billions of these yes/no machines can be linked together to produce perception, emotion, awareness, and the end product that these processes serve: adaptive behavior. Adaptive behavior produces animals that move smoothly through their environments, acting as if they were constantly asking themselves the question, "what should I be doing at this moment to maximize my lifetime fitness?"

How Nerve Cells Make Behavioral Decisions: The C-Start

One of the things that an animal must do to achieve fitness is to avoid being killed. If you're not alive, you can't reproduce.

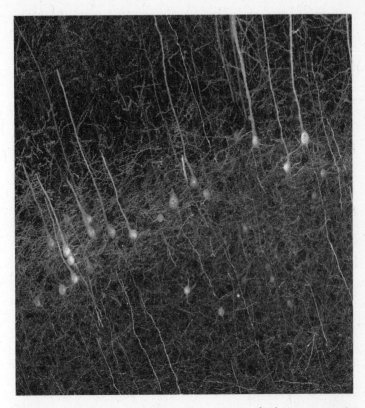

Figure 4.1 Neurons in the somatosensory cortex of a house mouse. In this area of the brain, information from the touch, pressure, stretch, temperature, and pain receptors is processed. Neuron cell bodies and their single axons are clearly visible, as is the dense web of synaptic connections on dendrites. Image courtesy of Kristina Micheva, Nicholas Weiler, Brad Busse, Nancy O'Rourke, Gordon Wang and Stephen J. Smith, Stanford University.

Picture a great blue heron standing in a shallow stream. The bird is hunting. Its long, powerful neck is cocked at a shallow angle to the water, and its head is tilted slightly to one side. There is no motion at all. To fish below, the heron appears like a tree branch fallen in the stream. When the bird sees a fish swim into range, it strikes, plunging its beak into the water in a motion so fast that you can't see it. You only perceive that a movement has occurred. The heron's beak may emerge holding a fish, but it may not. Failure to catch the fish does not occur because the bird's aim is poor. It occurs because fish can detect a heron's beak plunging through the water and quickly jump to the side. The jumping motion is called a *C-start*.

High-speed cinematography reveals that a C-start has two components. First, all the muscles on the side of the body opposite the heron's beak contract. This throws the fish's body into a C-shape, with the body wall rapidly bulging toward the heron's beak, pushing against the water and causing the fish to be propelled in the opposite direction. The fish then rapidly flips its tail, contracting the muscles on the other side of the body, to straighten out and scoot forward.

Like the plunge of the heron's beak, the C-start is blindingly fast. How fast is blindingly fast? The fish perceives the wave energy created by the heron's beak in one part of its ear, which contains hair cells that are very sensitive to acceleration. Two-hundredths of a second after the vibration reaches the fish's ear, the fish reaches maximum acceleration, at about fifty meters per second away from the stimulus.

The C-start is orchestrated by two large cells, called Mauthner neurons, in the fish's hindbrain. The Mauthner neurons lie close to the ear on both sides of the brain, and their huge axons cross the midline and descend in the spinal cord to make synapses with motor neurons on that side. Activity by the left Mauthner neuron causes all the muscles on the right side of the body to contract, and vice versa.

Axons from the ear hair cells branch to make two connections. The first is to excite interneurons that inhibit the Mauthner cell. These interneurons send axons to synapses on the Mauthner cell, and their activity creates IPSPs. This kind of nervous system design is called feed-forward inhibition. The second branch of the hair cell axons forms excitatory synapses directly on the Mauthner cell. Thus, at most times, when the fish's ear is detecting the kinds of vibration that are associated with water currents, a twig dropping into the stream, the fish's movement, or the movement of other fishes nearby, the IPSPs created by feed-forward inhibition outweigh the EPSPs and the Mauthner cell is silent. However, the rapid acceleration caused by the heron's plunging beak causes prolonged rapid volleys of action potentials from the ear hair cells, the EPSPs outweigh the feed-forward IPSPs, and the Mauthner cell fires an action potential.

The huge axon of a Mauthner cell excites motor neurons in the spinal cord on the opposite side of the body, but the axon also sends branches to interneurons that inhibit the motor neurons on the same side as the active Mauthner neuron. Another early branch of the Mauthner axon crosses to the other side of the brain and excites interneurons that inhibit the other Mauthner cell. Finally, another early branch excites interneurons that inhibit the active Mauthner cell. This design, in which a neuron turns itself off, is called *recurrent inhibition*.

Here we have a very clear explanation of the neural basis of a sign stimulus. Ethologists, by keen observation and simple experimentation, have shown that behavioral responses are often elicited by specific components of an overall stimulus. A heron's beak plunging toward a fish has a particular size, shape, color, and characteristic movement, but these are irrelevant to the fish. The fish's nervous system has evolved to detect that part of the heron's strike that is the most reliable indicator that a strike is indeed happening: the vibration created by the plunging beak. That vibration discriminates between a heron strike and all other events in the fish's life. The elements of nervous systems that are set up to make discriminations of this type are called *feature detectors*. The ethologists' truly amazing perception was that, using the concept of sign stimuli, they implied the existence of feature detectors before neuroscientists had described them.

The Mauthner-initiated C-start represents an instance in which we understand in complete detail how an animal evaluates the question, "what should I be doing at this moment to maximize my lifetime fitness?" The adaptive response to the heron's strike comes about by ignoring a lot of sensory information, by the arrangement of feed-forward inhibition and excitation of Mauthner cells that creates a feature detector for sudden vibration, by contralateral and recurrent Mauthner cell inhibition,

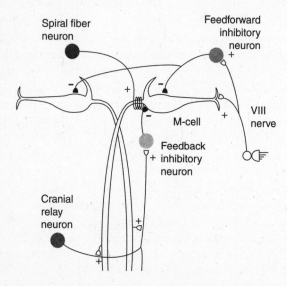

Figure 4.2 The basic Mauthner neuron circuitry, drawn by Minoru Koyama. The VIII nerve comprises the axons from the ear, where vibrations are detected. A plus sign indicates excitation and a minus sign inhibition. Figure courtesy of Joe Fetcho of Cornell University.

by fast traveling preemptive inhibition of muscles on the side of the strike, and, finally, by excitation of the contralateral muscles.

These design principles account for all animal behavior, and ultimately, the mechanistic basis for decision-making. The fly's decision to inflate her proboscis to dab at sugar on your table, your decision to rise and reach for the flyswatter, the fly's decision to jump and fly as your lethal strike started, all arise from a nervous system design that is fundamentally the same as the mechanism that caused the C-start. The main difference between the fish's decision and your decision lies in the number of interneurons that intervene between sensory input and motor output.

Crab Chewing

For obvious reasons, the Mauthner cells are referred to as *command neurons*. They exert complete, top-down control of a motor act. Although there are other examples of command neurons, there aren't many. Typically, motor output comes about as the result of interactions among several thousands of interneurons. To study the ways that neurons interact to create motor output, a good place to start is with the small networks of neurons that produce highly regular, patterned motor output: central pattern generators (CPG). CPG are responsible for rhythmic motor acts such as breathing, chewing, and walking, and they produce their motor commands on their own, without the need for timing messages from sensory neurons or other interneurons. CPG continue to produce their characteristic output even when removed from the animal and placed in a dish, so are useful in many kinds of experimental investigation.

Two famous CPG are those that direct the contraction of muscles in the stomachs of crabs, lobsters and their relatives. One – the gastric mill CPG – lies in the somatogastric ganglion, a small cluster of neurons inside a blood vessel where it passes across the top of the stomach. The gastric mill CPG creates the chewing movements of the fore-stomach, in which three hardened plates in the stomach wall are brought together to cut and grind food. In the lobster, the gastric mill CPG contains eleven neurons: ten motor neurons and one interneuron. The second – the pyloric CPG – contains fourteen motor neurons that generate the peristalsis-like movements of the second chamber of the stomach. In both CPG, the motor neurons send excitation signals to the muscles that they innervate, and inhibition signals to the other neurons in the CPG. The CPG neurons also fire spontaneous action potentials (due to specialized leaky $Na+$ channels) at different rates. Rhythmic chewing and peristalsis-like movements come about as emergent properties of small networks of spontaneously active, mutually inhibitory neurons. Neuroscientists have studied the crustacean somatogastric ganglion for over thirty years, and know the complete wiring diagram of each CPG as well as the individual membrane characteristics (density and types of ion channels) of each neuron. Thus, we have a clear understanding of how a crab manages to chew its food.

One of the surprises that emerged from this work was the discovery that a simple network of motor neurons can produce more than one pattern of neural output. For example, the gastric mill CPG can cause the three stomach teeth to come together in two distinct ways. Varied output from a simple network occurs because of another important cellular event: *modulation*. In addition to ion channels, nerve cell membranes have proteins embedded within them that bind with modulators delivered by other neurons. Modulators are neurotransmitter-like molecules. When they bind to the receptor proteins in the target cell membrane, they induce a cascade of cellular events inside the target neuron that alters the functional characteristics of the neuron's ion channels. This causes the neuron to respond differently to inputs from other neurons. Modulation alters the firing characteristics of individual cells, allowing a simple network of neurons to produce multiple, distinct outputs. In crabs and lobsters, over twenty neuromodulators affect the operation of the somatogastric ganglion. These come from neurons outside the ganglion and from the neurons that make up the CPGs.

The significance of the decades of work on the crustacean somatogastric ganglion was not that it allowed us to understand exactly how crabs manage to chew, but that it displayed the design features of a network of neurons. Crab chewing is a portal into understanding perception, thought, and motion.

Reward

Another class of chemical compounds that have modulating effects on neurons in the brain is hormones. Hormones are small molecules, produced by the endocrine glands, which enter circulation and travel throughout the body. The targets of any specific hormone are designated by cell receptors, proteins, embedded in the cell membrane, which bind to the hormone. Many neurons in the brain have hormone receptors and the binding of hormones to these receptors often causes profound shifts in behavioral priorities. As we saw in the previous chapter, animals evolve the ability to adjust their responses to stimuli when environmental contingencies make it adaptive to do so. One of the main mechanisms by which the adjustment of behavioral priorities occurs is by the modulating effects of hormones.

Consider a seasonally breeding temperate mammal such as the North American elk. Female elk come into estrus (showing the behavioral and physiological readiness to mate) in September, so males must be ready to mate and to compete for mating opportunities. Males achieve readiness through hormonal switching. After the summer solstice, the shortening day length initiates signaling in the male elk brain that causes the hypothalamus, a basic control center just above the roof of the mouth, to release a hormone called Gonadotropin Releasing Hormone (GnRH). GnRH is transported by a specialized vein to the anterior pituitary gland, located just below the hypothalamus. In response to GnRH, cells in the anterior pituitary gland release another hormone, Luteinizing Hormone (LH) into the circulation. LH binds to receptors on cells in the elk's testes and causes these cells to synthesize and release the hormone testosterone. Circulating testosterone has widespread effects. It induces the testes to grow and to produce sperm. It induces general muscle growth, and especially growth of the neck muscles. It partially suppresses the immune system, so that the male will have more energy available to compete. It causes the skin that covered and nourished the male's antlers as they grew to dry and shred off. It binds to receptors on neurons in the brain, inducing many changes in the male's behavior. The male begins to scent mark using his urine, spraying it on his chest and in the mud pits where he wallows. The male begins to emit squealing blasts, called bugles. He tries to monopolize access to groups of females, and becomes willing to fight with other males to maintain his monopoly. He becomes sensitive to the compounds in female urine that indicate the approach of estrus, and when he detects a female nearing estrus, he performs the actions that constitute courtship in elk. Finally, if successful in all these actions, he copulates with the females that accept him.

In this and in many other instances, hormones orchestrate integrated physiological and behavioral changes that allow animals to respond appropriately to their surroundings. For the most part, control operates along the same lines as for testosterone. Neural signals integrated in the hypothalamus cause the secretion of "releasing hormones" that are transported to the pituitary gland, where they induce or inhibit release of a pituitary hormone. The pituitary hormone enters circulation, binds to receptors on cells in the target endocrine gland, and induces the gland to release a hormone. Many hormonal effects on behavior prepare an animal to mate or to care for young. Hormones create the brain changes that induce a mother to form a bond to her young and they induce the formation of social bonds between mates in monogamous species.

Because hormones can have such strong effects on behavior, one way in which rapid evolutionary change in behavior can occur is by a change in the density of hormone receptors in the brain. This kind of change apparently was the basis for the evolutionary switch to monogamy in a small rodent, the prairie vole. There are many species of voles (also called field mice), and most are not monogamous. In most species, males seek out receptive females, mate with them if permitted, and then resume patrolling for other receptive females. In contrast, when a young prairie vole male finds an unmated female, the two copulate repeatedly and form a social bond. As a result of the bond, the male remains close to the female, guards her and the vicinity of the nest, and helps to care for the pups when they are born. This switch to monogamy in the prairie vole is in part due to a single mutation that results in a greater level of expression of the gene that makes the receptor for arginine vasopressin, a hormone made in the hypothalamus. A comparison of the brains of prairie voles to the brains of other species of vole shows a much greater density of arginine vasopressin receptors in the brain areas that are associated with reward.

Activity in the reward circuits of the brain is involved in learning, getting an animal to repeat behavior that is reliably associated with outcomes that lead to an increase in fitness. Rewards in effect tell animals that, when they eat, maintain a healthy body, and especially when they mate, that what they are doing is contributing to fitness. If a rat has an electrode placed into one of the reward areas of its brain and is given the opportunity to deliver electrical stimulation to that area by pressing a bar, the rat does much bar pressing. It becomes a bar press junkie, lying on its side so that its wrist is optimally poised over the bar to press a hundred times a minute for hours at a time, even when this means losing the opportunity to eat or drink. In humans, addictive drugs activate the reward circuits.

Brain Stimulation

Electrical stimulation of brains allows us to map function; to show where in the brain specific tasks are performed. Humans are good subjects for this kind of work. Brain stimulation is performed while the subjects are awake and humans can report what they experience when stimulation occurs. Brain stimulation of humans is done in the context of surgery for epilepsy or brain tumors, when the goal is to remove some brain tissue but to identify areas that should be spared.

Stimulation can induce basic physiological events such as blushing, nausea, changes in heart rate or breathing rate, changes in blood pressure, and sweating. It can also induce emotional states such as anxiety, fear, anger, sadness, and mirth, as well as motor outputs such as speech, changes in facial expression, eye movement, movement of limbs, and laughing or crying without feelings of happiness or sadness. It can induce the urge to move or the sensation that movement has occurred, visual, auditory, taste, and smell hallucinations, the recall of memories, the sensation of leaving the body, déjà vu, and impairments in reading, naming, or speaking.

These studies are useful in that they show where emotions, sensations, and movements are generated, but they don't take us very far into the understanding of the actual neural mechanisms involved. Greater understanding is made possible by stimulation experiments on animals. The animals can't report what they are feeling, but it is possible to study how motor output is produced in a more controlled, systematic way than is ethically permitted in humans. Recent work on monkeys has shown that there are areas of the so-called pre-motor cortex where motion goals seem to be encoded. For example, stimulation in one area results in a monkey bringing its hand to the center of its chest, irrespective of the initial position of the hand. If the hand is below the chest, one set of muscles must be contracted, but if the hand is above the chest, another set is involved. So the "intent" center works through an output area that incorporates feedback from the current position of the limb.

Nevertheless, there is a somewhat frustrating disconnect between the studies like those of the crustacean somatogastric ganglion, where we understand in almost complete detail how a network of neurons produces motor output, and studies of movement that is produced by the brains of flies, monkeys, or humans. Most neuroscientists would agree that studies like those of the somatogastric ganglion are a portal to the understanding of more complex output of nervous systems. However, there is still a long way to go. As Professor Michael Dickinson an integrative neurobiologist and the Ben Hall Endowed Chair at the University of Washington put it: "Fortunately, many challenging problems remain, otherwise, engineers and toy makers would have already littered the world with tiny mechanical flies."

5

Instinct

If we step outside on a summer morning almost anywhere in the world, we hear birdsong. If we live in a natural or even a semi-natural area, what we hear is better described as a cacophony. Birdsong is loud and it is repetitive.

Male birds perform most of the singing. Each male claims an area of space from which he will attempt to exclude all other birds except his mate. This is his territory. The territory contains the site where the male and his mate build their nest, and often enough surrounding habitat to provide the food that the male, his mate, and their brood will need.

Male birds defend their territories by patrolling the boundaries, threatening and chasing away intruders, and occasionally by entering into knockdown, fall-to-the-ground-and-pummel-your-opponent fights. But patrolling, chasing, and fighting are costly activities. Although territoriality will persist, evolutionarily, as long as the overall cost is less than the benefit, any individuals that are able to shrink that cost will have that much more resources to put into their offspring, and hence a higher fitness. One way that a male bird can reduce the cost of territorial defense is to advertise that the territory is occupied. He can sing.

Birdsong is loud and repetitive, but it also seems difficult to produce. The difficulty in production is not immediately easy to understand. It seems reasonable to think that a bird could advertise his presence on a territory perfectly well with the vocal equivalent of a vuvuzela – a loud, single note. This is not what birds do. Instead, they produce loud, multi-note, complex sounds that genuinely qualify as song. The sounds that they produce

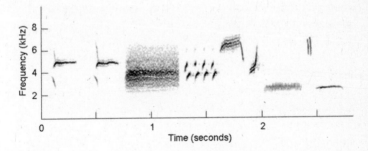

Figure 5.1 Sonogram of one song type produced by a song sparrow male. Figure courtesy of Mike Beecher at the University of Washington.

usually represent virtuoso performances that far exceed the abilities of any human vocal performer.

Figure 5.1 shows a sonogram (a visual representation of a sound) of one of the songs of a male song sparrow. Time is represented on the *x* axis, and sound frequency is represented on the *y* axis. Amplitude at any time and frequency is represented by the amount of black. In the 1940s, before this technology existed, Donald Griffin struggled with ways to visualize and measure the echolocating clicks of bats. Now, anyone can easily download and use the software that makes these visual representations of sound.

Listening to the two-second song depicted in Figure 5.1, you would hear two distinct introductory notes, a breathy buzz, four punched, staccato notes, a high tone followed by a grace note to a low buzz, and another high grace note before a low whistle. At normal speed, the inexperienced listener would find it difficult to discern these distinct song elements because they are packed together so closely. The human listener also would not hear the steep frequency sweeps (a vocalist would recognize these as impossible glissandos) in the introductory notes and in the high tone and grace note that follow the staccato passage. Finally, in the two introductory notes, while the bird is producing the loud clear tones that start with a short upward frequency sweep, it is simultaneously producing a lower note that is a downward frequency sweep! Birds are capable of vocal feats like this because the syrinx, where sound is produced, has vibrating membranes on the right and left sides with separate muscular control and innervation.

Why birdsong embodies such striking virtuosity is not known, but the question is under investigation in quite a few laboratories. Whatever the evolutionary history and current function(s) of bird-song, we know that it is a highly complex, skilled motor performance that must involve precise control of the muscles that control breathing, the syrinx, the throat, the trachea, and the beak.

How Does that Song Go?

Peter Marler, who provided a key conceptual entry into the understanding of birdsong, began to study white-crowned sparrows in the bay area of San Francisco in 1957, when he took a faculty position at Berkeley. Extending work that he had done on some European birds for his doctorate at Cambridge, Marler sought to explain the existence of local song dialects in the sparrows. "Local song dialect" means that if we cross the Golden Gate Bridge and record the song of white-crowned sparrows in Marin County, their songs, although species-typical, consistently sound a bit different from song recorded in Palo Alto. Birds in Berkeley might have another, distinct dialect. Over relatively short distances (for animals as mobile as birds), there is local, slight, but distinct differentiation of their song. Marler's approach to figuring out why dialects persisted involved describing how song appeared to develop in nature, and then conducting a series of laboratory experiments, in which he intervened at different points in that development.

In nature, a white-crowned sparrow male hatches and is fed by his parents until he fledges (is able to fly away from the nest).

During this time, he frequently hears his father singing, but he himself is silent, except for the begging calls that he emits to elicit parental feeding. Sometime well after fledging, at about 150 days of age, the young male begins to sing. His first songs are rudimentary, rather poor copies of white-crowned sparrow song, but he continues to sing, and over the next fifty days, his song improves until it sounds just like his father's. This period of practice and improvement is called subsong, or plastic song.

Marler's experimental approach was to take newly hatched nestlings from nature and to raise them in a laboratory, where he could record all the sounds that the birds produced and could control the sounds to which they were exposed. An extensive series of experiments established several facts about the way that song develops in white-crowned sparrows. For example, if birds are taken from the nest in the first week after hatching and raised in auditory isolation, they enter subsong at the appropriate age, but the song does not improve to the adult type. Marler obtained the same results when the young birds were exposed only to the songs of another species, the song sparrow. In other words, if a young white-crowned sparrow hears only the songs of another species, it is as if he has heard nothing. Further experiments showed that, to develop normal song, the young male must hear white-crowned sparrow song sometime between ten and fifty days after hatching. This is an example of what is called a sensitive period. There is a range of ages between which the brain may be modified by experience. If the necessary experience falls outside the sensitive period, the brain modification does not occur.

The young bird is on track to develop normally if it hears white-crowned sparrow song during the sensitive period for song learning. Then, about a hundred days later, the bird enters subsong. Marler hypothesized that, during subsong, the bird must be comparing the sounds that it produces to the songs that it memorized during the sensitive period, trying for a better and better match. To test this hypothesis, Marler deafened birds before they started subsong. The results were as predicted. Birds that could not hear themselves did not improve. A bird must be able to hear itself singing during subsong to alter the motor program to that of adult song. Marler also deafened birds after song development was complete, and showed that song was unaffected. White-crowned sparrow males must be able to hear themselves while they are in the self-tutoring phase of subsong, but after that, the motor program is set in its final form (Marler's term was "crystalized"), and sensory feedback is irrelevant.

To recapitulate, a white-crowned sparrow male hatches. From its nest, it hears its father singing, the songs of neighboring white-crowned sparrow males, the songs of other bird species, crickets chirping, frogs calling, perhaps automobile horns and human voices. The bird has an instinctive feature detector that allows it to recognize white-crowned sparrow song. When the feature detector is activated, a memorization program starts, and the young bird stores a detailed accurate copy of adult song. Three to four months later, the bird starts to sing, and it compares the songs that it produces to the memorized song, modifying motor output until there is a perfect match. Marler had found the explanation for song dialects. Local dialects exist because of the memorization-based mechanism by which song develops.

Marler discovered a program of behavioral development that is nearly ubiquitous in songbirds and is now referred to as sensori-motor song development. In songbirds, the basic arrangement of a memorization phase followed somewhat later by a practice phase is the norm, but there is substantial variation in several aspects of the learning program. The duration of the sensitive period for memorization varies from a fairly narrow window (as is true for white-crowned sparrows), to the first entire year of life, to a lifetime ability to learn. Some researchers categorize bird species as either closed-ended or open-ended learners. At the extreme open end are bird species, such as mockingbirds or starlings, which are called mimics. Seemingly throughout life, individuals of these species are able to memorize and then produce accurate copies of sounds. A starling living with humans repeated the question, "Does Hammacher Schlemmer have a toll-free number?" after hearing a human say it once. Among non-mimetic species, there is still quite a bit of variation in how many songs are learned, how accurately the memorized songs are copied, and the stringency of the feature detector that turns on song-memorization.

Instinctive and Learned

Is birdsong instinctive, or is it learned? Is it nature or nurture? Obviously, it is purely neither. It is both. Birdsong is not completely instinctive, because the young bird must learn by memorization exactly how its songs should sound, and must learn to produce these sounds by practicing. Birdsong is not completely learned, because the bird has a feature detector that identifies what it should memorize, and it has a preset developmental schedule that determines when the sensitive period for learning will open, when it will close, and when the period of subsong will start. Birdsong develops through an interaction between instinctive and learned components. It is a form of programmed learning. When we understand how song develops, to ask whether song is instinctive or learned begins to seem irrelevant.

The evolutionary biologist Ernst Mayr offered a better way to think about the development of any motor act. Mayr noted that, instead of a dichotomy (nature–nurture), there was a continuum between what he called closed and open motor programs. At the closed end are motor acts that do not develop; the motor pattern exists in its complete form when the animal is born. Behavioral acts at this end of the continuum are akin to those that ethologists use to illustrate the concept of instinct. Immediately after hatching, a nest parasitic cuckoo chick pushes other eggs and chicks out of the nest. A red kangaroo, like all marsupials, is born as a tiny embryo. The embryo must travel from where it emerges, the opening of the vagina, upward to the mother's pouch, where it will find a nipple and so, nourished by milk, be able to complete the remainder of its development. When a red kangaroo is born, its hind limbs are rudimentary but its front limbs are functional and muscular. The embryo swings its trunk to the left to advance the right forefoot, then to the right to advance the left forefoot, thrashing from side to side like a salamander to move upward to the lip of the pouch. The mother does not assist. The embryo makes this epic trek on its own, so the thrashing motor program must be ready at the moment of birth. A human infant, like all mammal infants, when touched on the lips by the mother's nipple, opens the mouth, closes the lips around the nipple, and initiates the complex motor program of sucking.

A bit further along the continuum is behavior such as birdsong, with its programmed memorization and practice. Still further along is locomotion in the fancy movers, the birds and mammals, in which central pattern generators control basic locomotion but fine control of movement develops through extensive, performance-based learning. In birds and mammals, the cerebellum, a part of the brain where fine tuning of motor commands occurs, is large. In mammals – and probably also in birds – there is a postnatal sensitive period during which performance-based loss or retention of synapses in the cerebellum occurs. There is an initial overproduction of synapses. Then, during the sensitive period, the synapses that are active during a movement are retained, and the synapses that are not active tend to be lost. In mammals, this sensitive period coincides with the ages during which young individuals express the extravagant leaps, twists, cavorting, running, and mock fighting collectively known as play. Also somewhere in this neighborhood of the continuum is human speech, which has some lifelong learned components, but which also shows striking developmental similarity to birdsong. Like birdsong, human speech development involves an initial memorization period aided by feature detectors that instruct the infant what to memorize. Also like birdsong, there is a sensorimotor phase, babbling, in which the infant appears to compare the sounds it produces to what it has memorized. During babbling, speech does not improve if the infant cannot hear. Finally, like birdsong, human speech appears to have specific sensitive periods. Human languages share many phonemes (the individual motor acts of speech) but not all. If a child is not exposed to a phoneme before puberty it will not be able to hear the phoneme as distinct from others, and will not be able to produce it correctly. Immigrants who arrive before puberty end up sounding like native speakers, but those who arrive later always have a distinct foreign accent.

To summarize, it is not useful to label a motor pattern as either instinctive or learned. That kind of labeling closes the door to the investigation of how motor patterns develop; investigations that often reveal a blend of instinctive and learned components. Some motor patterns genuinely are closed. They exist in their final form when the animal hatches or is born. But in many instances, and really to an unknown extent, motor patterns become modified by environmental input to which the animal's nervous system has a programmed sensitivity.

Instinctive Responses to Stimuli?

So far, our discussion has focused on motor patterns, or motor output. Now we must move on to consider stimulus and response. Animals' responses to stimuli might be either instinctive or learned. However, when we study the development of such responses, we see the same patterns that exist for the development of motor patterns. Some responses are indeed "hard wired" and unvarying, but many develop through an interesting blend of instinctive and learned components.

How Bats Identify Water

When bats are foraging, plucking insects out of the night sky, they also need to drink, which they do by flying low across the surface of a river or lake and dropping the jaw to scoop up water. In the dark, how can an animal with very weak eyesight identify a body of water? This was a question that researchers at the Max-Planck Institute for Ornithology recently sought to answer. The researchers proposed that bats could use echolocation, because the echo characteristics of a smooth surface are distinct and different from the characteristics of any sort of textured surface.

The researchers reasoned that, if bats use the echo properties of water to identify water, they should attempt to drink from any surface that had those properties. In nature, no other surface has these properties, but in the laboratory, surfaces like this can be presented to bats. Using metal, wood, and plastic, the researchers made smooth "water" surfaces and textured surfaces. In a flight room, they placed one sheet of smooth and one sheet of textured material on the sandy floor, and released a thirsty bat. The bats attempted to drink from the smooth surfaces and did not attempt to drink from the textured surfaces. In five minute-long trials, bats attempted to drink from the smooth surfaces about fifty to a hundred times, and continued to do so even after they had landed on those surfaces. The researchers then placed the smooth surfaces on a table in the flight room, to discover if water that appeared to be hovering in mid-air would be less attractive. Even though the bats flew underneath the table (perceptually, underwater) they continued to attempt to drink from the smooth surface. The researchers ran these tests on fifteen bat species and got the same results, suggesting that all echolocating bats identify water using this feature detector.

Finally, the researchers captured young bats, and their mothers, from a cave before the juveniles could fly, and raised them in the laboratory. When the juveniles could fly, they were tested. These animals had never flown over water, but nevertheless, they tried to drink from the smooth surfaces on their first flights in the test room. Bats are born with a feature detector to identify water and use it, unmodified, throughout life.

Learning Where to Peck

One of the classic studies in ethology concerns the investigation of the signaling that coordinates feeding by parents of newly hatched chicks in several gull species. In nature, a gull chick hatches and then spends an hour or so getting dry, its downy coat fluffing up to provide insulation. One or both parents are present to warm the chick if need be, and to shield it from gulls in neighboring territories, who would jump at the chance to devour an unprotected chick. After the chick is dry and standing, the parent typically regurgitates some food (such as half-digested fish) into the nest, breaks off a morsel, and holds it in the tip of its beak, held at a downward angle toward the chick. The chick pecks at the parent's beak, eventually strikes the food, and has its first meal. On this and on subsequent occasions, the chick may peck first at the parent's beak, inducing the parent to regurgitate.

Niko Tinbergen directed the first systematic studies of the pecking response of gull chicks. Tinbergen reasoned that a gull chick could not hatch with a picture of the parent's head in its brain. Therefore, there must be a simple set of features about the parent's head that were sufficient to elicit pecking. The features had to be sufficient to provide a target, because the chick reliably pecks near the tip of the parent's beak.

Tinbergen's initial studies, conducted in the Netherlands shortly after the end of World War II, were on herring gulls. Like many other closely related gull species, adult herring gulls have a light-colored head and a yellow beak that has a conspicuous, roughly circular, red spot (the *gonydeal spot*), on the lower part, near the tip. To figure out the key stimulus elements that elicited chick pecking, Tinbergen and his student helpers searched through the gull colony for damp or newly dried chicks that had not yet had a feeding experience with a parent. They brought these chicks to a nearby sheltered spot and presented them with models of an adult gull's head. The painted models, cut from heavy paper, were attached to a short rod that the experimenter used to move the model head back and forth in front of the chick. During each model presentation, the experimenters counted the number of pecks that the chick aimed at the model. Tinbergen and his coworkers systematically altered the color, size, and shape of the models to discover what aspects of the overall stimulus elicited the chicks' pecking. They also made three-dimensional models and in a few instances presented the actual head of a dead gull. These tests revealed that three-dimensional depth of the stimulus was not important; chicks responded just as strongly to the paper cutouts.

Tinbergen's main results and the nature of the data from which he drew his conclusions are shown in Figure 5.2. Note that the shape of the beak alone is as effective as the shape of an entire head. Chick pecking is elicited most strongly by a moving narrow rectangle that has a dot of contrasting color near the tip. Chicks hatch, not with a picture of the parents in their heads, but with a feature detector. The behavior of the parents and the appearance of the parental beak turn on the feature detector. When Tinbergen homed in on this constellation of stimulus characteristics (to ethologists, the effective constellation is called a sign stimulus), he was able to construct an exaggerated version, a very thin strip with contrasting color near the tip, that elicited more pecking than a complete model of a parent's head. He had constructed what ethologists refer to as a *supernormal stimulus*.

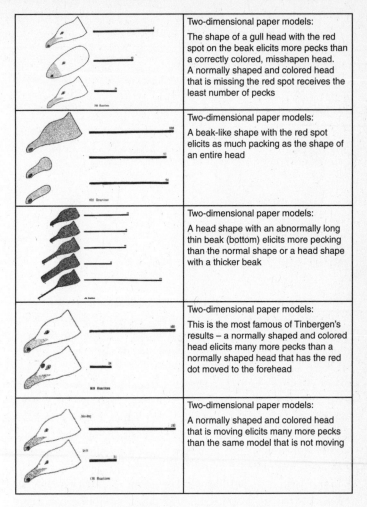

	Two-dimensional paper models: The shape of a gull head with the red spot on the beak elicits more pecks than a correctly colored, misshapen head. A normally shaped and colored head that is missing the red spot receives the least number of pecks
	Two-dimensional paper models: A beak-like shape with the red spot elicits as much packing as the shape of an entire head
	Two-dimensional paper models: A head shape with an abnormally long thin beak (bottom) elicits more pecking than the normal shape or a head shape with a thicker beak
	Two-dimensional paper models: This is the most famous of Tinbergen's results – a normally shaped and colored head elicits many more pecks than a normally shaped head that has the red dot moved to the forehead
	Two-dimensional paper models: A normally shaped and colored head that is moving elicits many more pecks than the same model that is not moving

Figure 5.2 Some of the models that Tinbergen and his coworkers used to study the stimulus characteristics that elicited pecking by newly hatched herring gull chicks. The length of the bar beside each model represents the strength of the chick's pecking response to the model. Figures reproduced from: Tinbergen & Perdeck (1950). 'On the Stimulus Situation Releasing the Begging Response in the Newly Hatched Herring Gull Chick' (*Larus argentatus Pont.*). *Behaviour* 3(1), 1–39.

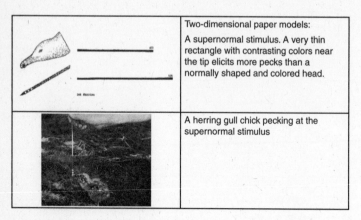

| | Two-dimensional paper models: A supernormal stimulus. A very thin rectangle with contrasting colors near the tip elicits more pecks than a normally shaped and colored head. |
| | A herring gull chick pecking at the supernormal stimulus |

Figure 5.2 (continued)

Experimenters can construct supernormal stimuli because animals commonly rely on feature detectors to guide their behavior.

Tinbergen, with the lucidity that characterized his perception and thinking, correctly depicted the signaling system between the parent and chick. Chicks hatch with an instinctive feature detector that guides their pecking. The feature detector is turned on most strongly by a moving narrow rectangle that has a contrasting spot of color near the tip. Chick pecking induces the parent to regurgitate food. Recently, researchers in the Netherlands replicated Tinbergen's experiments, using more stringently controlled methods, and confirmed his results.

Tinbergen's conclusions were incorrect in only two respects, one somewhat trivial and one more important. The trivial error was that Tinbergen concluded that the relative position of the red spot on the parent's beak was important. Figure 5.2 shows that the model with the red spot on the parent's forehead elicits less pecking than the model with the red spot in the normal position. Tinbergen's interpretation of this result was that the chick hatched with a kind of rudimentary image in its brain, which included the correct position of the red spot. Subsequent work by an American, Jack Hailman, proved that chicks actually attended to the angular velocity of the red dot, not its position with respect to the other features of a parent's head.

The second adjustment to Tinbergen's conclusions that Hailman made was more significant. Hailman studied the pecking responses of newly hatched chicks and the pecking responses of chicks that had been in the nest for a few days. He proved that after a chick has fed from a parent for a day or so, it has lost the feature detector-driven response. It no longer will peck at an isolated beak, and it also discriminates between its own and other species. The chick has learned to recognize the parent. The sign stimulus that Tinbergen worked hard to identify is important for directing the chick's first pecking responses to the parent, but only for the first day or two after hatching. When the parent delivers food and the chick can see who is feeding it, the chick begins to memorize the exact appearance of the parent. Feature detector-driven response to the parent is replaced by memorized facial recognition of the parent. Responses to stimuli, like motor patterns such as birdsong, often develop from a mixture of instinctive and learned components.

A human baby may smile before it is born, but not because it is happily anticipating the painful squeezing that it is about to experience. The baby smiles because its nervous system is rehearsing a motor program that will be used soon after birth. After it is born, a baby will smile in response to a face-like object. This smiling response helps to establish the bond that the parent develops with the child, even though a young infant cannot see clearly enough to discriminate among faces. The baby will smile at a cardboard face-shaped disc, or at a disc with eyes, nose and mouth in scrambled orientations. Later, like gull chicks, the human infant replaces this feature detector-driven response with a response that is based on memorization of the facial characteristics of its parents.

6

Learning

In the Stanley Basin of central Idaho, the White Cloud, the Boulder, and the Sawtooth Ranges form a rugged bowl around beautiful Redfish Lake. The name of the lake derives from the once great numbers of sockeye salmon that came there to spawn. The males have red backs when they are in breeding condition, and reportedly the surface of the lake had a reddish cast during spawning season. To get to the lake, the fish swam about 1,600 kilometers and gained about 2,000 meters in elevation from the mouth of the Columbia River. After spawning, all the adults died, leaving billions of developing eggs in the gravel.

Salmon are *anadromous*: they hatch in fresh water, migrate to the ocean or an inland sea, and return to fresh water to spawn. All make trips similar to the fish of Redfish Lake, although all other trips from the ocean are shorter. All salmon show strong "home stream fidelity;" the adults return to spawn where they hatched.

Consider how, remarkable the homing performance of salmon is. A tiny smolt, about eight to ten centimeters long, drifts into a river drainage and descends to the ocean. Several years later, this individual, which may have been far out to sea, returns to the river mouth from which it emerged, and follows in reverse the turns it made long ago, until it reaches the stream where it hatched. How do salmon accomplish this extraordinary feat? Arthur Hasler of the University of Wisconsin discovered most of the answer to this question in the mid-twentieth century. Hasler hypothesized that salmon find their way home by smell. Because each stream or lake has its own bedrock, its own surrounding soil, and its own surrounding plant community, each has its own specific cocktail of dissolved chemical compounds. Hasler realized of early, rapid, irreversible learning that allows young waterfowl to follow the mother and later to express sexual interest in the correct species. Because Lorenz was the "parent" of several clutches of geese (that is, he was the only moving object present when they hatched) he found himself followed by young goslings and later courted by these individuals when they became sexually mature. Lorenz used the term *imprinting* to denote this early learning because the effect occurs rapidly and seems to be irreversibly stamped on the animal's brain. Later research revealed that there were two distinct processes involved. The first, filial imprinting, occurs in a short sensitive period just after hatching. The young duck or goose memorizes the appearance of the first moving object that it sees, and forms an attachment to that object. When the object moves away, the juvenile follows. If the juvenile is visually separated from the object, it emits distress calls. In nature, that first moving object seen is the mother. If inquisitive humans intervene, that object may be Konrad Lorenz, or a toy train, and the hatchling will follow those. In most species, sexual imprinting occurs during a later and less sharply defined sensitive period. Nevertheless, it can be potent. Sexual imprinting resulted in the failure of the Grays Lake, Idaho whooping crane recovery effort, begun in 1975 and stopped in 1989. Wildlife biologists wanted to augment the number of whooping cranes, an endangered species. Like other cranes and related birds, this species lays a clutch of two eggs, but only one juvenile survives. The biologists charged with the recovery of whooping cranes reasoned that they could take this doomed egg from whooping crane nests, and have it raised by foster parents. They placed the eggs in the nests of the closely related sandhill crane in the Grey's Lake Wildlife Refuge in southeastern Idaho. The sandhill crane parents accepted these eggs, raised the young, and showed them the migratory path to the wintering area in New Mexico. As adults, these cross-fostered whooping cranes showed no sexual interest in other whooping cranes, only in sandhill cranes.

They had sexually imprinted on the model provided by their foster parents.

Social Bonds

Narrow sensitive period learning also occurs as a way to create exclusive social bonds, the basis for monogamy. Prairie voles, by copulating, excite reward circuits that guarantee that a bond is formed with the partner. Exclusive social bonds are also the basis for the adaptive provision of maternal care. In mammals that form large groups, such as most hoofed animals, seals, and bats, a sensitive period that occurs immediately after birth allows the mother to memorize the odor of her young. Learning of the odor signature of young occurs in about the first hour after birth, while the mother is avidly licking the young. If the mother's ability to smell is experimentally blocked during the short sensitive period, the mother does not memorize the odor and will not accept the young. This specialized, rigid, and quite specific form of learning exists to give the mother an infallible ability to identify her own young. This ability is adaptive because milk is very expensive to produce. To maximize her lifetime fitness, a mother should deliver this precious resource only to her own offspring.

The unifying theme of all these examples is that animals are able to extract useful information from the environment and use it to guide or adjust their future behavior. This characteristic of animal brains is called *learning*. When a salmon smolt memorizes the odor of its home stream, when a duckling identifies and bonds to its mother, when social bonds are formed between mates, or between mothers and young, the learning is programmed to operate at the appropriate time and context, and then turned off. Other forms of learning are not governed by sensitive periods and appear to operate throughout life. However, all learning is programmed insofar as what animals learn, the context in which learning occurs, and how rapidly learning and forgetting happen, are preset. There is no *tabula rasa*. Brains can only learn because they are built with specific learning instructions.

What to Eat and What Not to Eat

The Norway rat, a source of human misery and disease for centuries, originated in China. In nature, this species lives in complex burrow systems that contain related females and young, guarded by a single male. The species is a generalist feeder; it consumes many kinds of foods, including fruits, seeds, plant leaves and stems, and small animals. These two characteristics allowed rats to expand into a new niche, living alongside humans. The spaces beneath human dwellings or within the walls resemble burrows, and generalist feeders have no trouble adapting to the food sources offered by human households and agriculture. Rats lived alongside humans in China, and reached Europe in ships when trade between the East and the West began.

Generalist feeders like rats have feeding traits that allow them to continually add new items to the diet. When a foraging individual finds something new that might be food, it eats a tiny amount, enough to form a memory of the novel smell and taste. If, in the next few hours, the rat becomes nauseous, it develops a strong aversion to the smell and taste that it memorized. This specialized ability is called one-trial food avoidance learning. The phenomenon was first described in laboratory strains of the Norway rat (the characteristics that allowed rats to live alongside humans have allowed humans to domesticate them as laboratory animals). Researchers gave laboratory rats a novel, palatable food and then exposed them to moderate doses of X-radiation. The nausea produced by radiation sickness induced avoidance of the novel food.

Rats are also able to acquire preferences for novel foods from each other. When a forager returns to its burrow, its extended family group of rats greet it in short bouts of nose-to-nose sniffing. In the laboratory, this sniffing allows individuals to detect the scents of new food on the breath of the returning forager, and causes them to develop a new preference for food bearing that scent. This social acquisition of a food preference for a novel food occurs rapidly and its effect is robust. The context in which this social learning operates in nature is unknown, but it is reasonable to surmise that social acquisition of food preferences, like one-trial food avoidance learning, is part of the learning package that supports a generalist feeding habit.

Neophilic Ravens

Another spectacular example of timed, programmed learning that exists to promote optimal diet breadth occurs in common ravens. Ravens are one of the most pronounced habitat and diet generalists among perching birds. The species occurs from arctic tundra, to mountains, to seashores, to deserts and even to human cities, and, across these habitats, individuals consume a huge variety of foods. However, adult ravens, unlike adult rats, are neophobic. They avoid, and appear to be afraid of, food items that they have not previously encountered. Bernd Heinrich, who has studied ravens more than any other biologist, designed an experiment to investigate this paradox. How can ravens be generalists if adults are afraid of new foods? The answer, Heinrich discovered, lay in the demeanor of juvenile ravens.

Unlike adult ravens, juvenile ravens have a profound neophilia, which Heinrich demonstrated in a very compelling way. Working with four hand-reared young ravens, Heinrich began each morning session by feeding the young birds as much meat as they wanted. Then he opened the door to the outdoor aviary and led the birds into an arm-like extension, which had a natural floor of grass and dead leaves. Initially, Heinrich kept track of which objects the young birds picked up and manipulated in their beaks, and which they ate. In subsequent trials, he added new items, some edible and some inedible. The birds initially picked up every class of object in the arena. Then, when new items were added, the birds instantly switched from what they had previously investigated to the new items. If the new items were edible, the birds developed a preference for them, but continued to investigate new items. Their ability to find new items, even tiny ones, was extraordinary. At later ages, the same birds showed aversion to novel items. Thus, ravens are programmed to show an intense attraction to novel objects during what is likely a sensitive period in development, at a time when they would be following their parents after leaving the nest. During this time, they learn to discriminate between the edible and the inedible, and they retain the developed preferences for life. The abilities of young ravens to remember every class of object that they have already handled and consequently to instantly recognize novelty is remarkable.

Faces

Another example of specific, rapid, directed learning occurs in humans: the almost instantaneous formation of a memory for a face. The ability of humans to perform this task rivals speech processing and production for the sheer complexity of the information storage and retrieval involved. Humans can remember thousands of individual faces and can recognize a face seen twenty years before and altered by age. Like many forms of learning, facial learning occurs automatically, without the need for any specific reward or punishment. Nor is any conscious effort required. Facial learning is a program that starts running in the background whenever we perceive a face. Facial learning occurs in a dedicated part of the temporal lobe, the *fusiform gyrus*.

Locations

Another kind of learning that, like facial learning, is programmed to run automatically, without any associated punishment or reward, supports the ability of animals to remember locations and to select efficient travel routes. Like facial learning, spatial learning has its own dedicated brain area, the *hippocampus*. The hippocampus performs functions other than spatial memory, but spatial memory is centered there. In the hippocampus are neurons called place cells. Each place cell is assigned to a physical location in the environment that the animal has visited. When the animal returns to that location, the associated place cell becomes active. The automatic assigning of place cells in the hippocampus to physical locations in the environment allows animals to develop what has been called a *cognitive map*: a mental representation of points in space and their physical proximity to each other. Several comparative studies have shown that species that perform complex spatial memory tasks have a hippocampus that is relatively larger than the hippocampus of closely related species that do not, and that the size difference is associated with longer retention of spatial memories. Even within a species, the hippocampus can increase in size in response to increased demands for spatial memory. London taxi drivers who have worked for several years have a larger posterior hippocampus than control subjects who do not need to memorize and use a detailed spatial map for their work.

Several species of birds and mammals that eat seasonally abundant seeds have evolved a way to harvest and store the seeds for use during months when the seeds are not available. These so-called "hoarding species" use two distinct hoarding strategies. *Larder hoarders*, such as golden hamsters or acorn woodpeckers, place all the seeds that they harvest in a single, defended, location. This strategy is simple and easy to carry out, but is only as good as the ability of an individual animal to defend the larder from thieves. When a larder can't be defended, or when the cost of defending is excessive, species use another strategy, *scatter hoarding*. Scatter hoarders, such as grey squirrels, many kangaroo rats, and several species of jays, harvest many seeds but disperse them into dozens to thousands of small isolated caches. A scatter hoarder avoids the risk of losing the entire larder to a thief, but of course must be able to remember all or most of the cache locations. Although it may seem improbable that an animal can remember, in some way, thousands of individual locations in space, scatter hoarders do just that.

Clark's nutcracker, thus called because the Lewis and Clark Corps of Discovery first described it scientifically, is a stout jay of western North America. The species is specialized to exploit the seeds of pinyon pines. In fall, individual birds harvest the large seeds, cram them into a specialized pouch in the throat, and transport them to high elevation spots. They may fly more than twenty kilometers to reach these sites, where they cache the seeds. Individuals use their powerful beak to thrust seeds into the ground, creating caches of one to fourteen seeds. In a year when the pines are productive, a single bird caches twenty to thirty thousand seeds in about seven thousand dispersed locations. In winter and spring, the birds return to their caches to feed themselves and their nestlings. Thus, birds must often dig through snow to recover the seeds from a cache. Because snow obscures all local landmarks, it is obvious that the birds have some kind of mental representation of the spatial array of their own caches. Because each cache is created in about thirty seconds, these spatial memories are formed rapidly. Like ravens, Clark's nutcrackers pose a huge challenge to neuroscientists: how can a bird with a brain the size of a kidney bean store such a massive amount of information?

Habituation

Some forms of learning have a more general character. They operate throughout an animal's life because they are involved in adjustment to a persistently changing environment. Although nature may seem constant (year after year, you can visit a natural area and detect no obvious changes), it is actually in constant flux.

carried out in Pavlov's laboratory, a dog, restrained by a harness, stood on a table. Days or weeks earlier, Pavlov had cut through the dog's cheeks to insert and to sew into place a cannula (glass tube) that provided access to each side of the dog's mouth. The physiological reflex that Pavlov studied was the dog's release of saliva into the mouth in response to food. Pavlov could blow a puff of meat powder through one cannula and measure the volume of saliva that dripped from the other into a graduated cylinder. Pavlov referred to the meat powder as the *unconditioned stimulus* and to the production of saliva as the *unconditioned response*. Then Pavlov began to introduce another stimulus, such as a bell ringing, at the same time as the puff of meat powder. After several trials, in which the stimuli appeared simultaneously or nearly so, it became possible to induce the production of saliva by presentation of the bell alone. The dog would salivate even though no meat powder was in the mouth. Pavlov referred to the ringing bell as the *conditioned stimulus*, and to salivation in response to the bell alone as the *conditioned response*. The entire procedure is now known as Pavlovian conditioning, or classical conditioning.

An important characteristic of Pavlovian conditioning is that the dog (and any other animal that can be trained in this way, such as fruit flies) can't be permanently fooled. If the bell (conditioned stimulus) is presented alone, without the meat powder in the mouth (unconditioned stimulus) for several trials in a row, the production of saliva in response to the bell (conditioned response) begins to wane and eventually goes away. Pavlov referred to this unlearning of the conditioned response as *extinction*. Any stimulus that the dog can perceive will work as a conditioned stimulus. Punishing unconditioned stimuli, such as sources of pain, are as effective as rewarding unconditioned stimuli, such as food. The existence of Pavlovian conditioning means that animals are able to temporarily assign previously insignificant events (such as a bell ringing) that are reliably associated with pain or reward as indicative of pain or reward and to respond appropriately.

When the association between the stimulus and the fitness-related event goes away, the animal begins to forget the association.

The mechanisms that produce Pavlovian conditioning also produce what is called *instrumental learning*. In Pavlovian conditioning, the animal begins to associate the conditioned stimulus (bell) with the unconditioned stimulus (food). In instrumental learning, the animal begins to associate some behavioral act that it performs with the unconditioned stimulus. For example, imagine a psychologist builds a cage for a rat that contains a metal bar on a spring-loaded hinge with a small trough below. When the rat presses the bar, a food pellet drops into the trough. A rat will quickly develop a bar pressing response. Or the psychologist could build a cage in which half of the floor is made of metal bars that can carry an electric current which will shock the rat. If a conditioned stimulus such as a light or a bell is paired with the shock, the rat will quickly learn to run to the other side of the cage when the conditioned stimulus is presented. Like responses built up by Pavlovian conditioning, responses built up by instrumental learning also show extinction when the pairing between the behavioral act and the punishment or reward is removed. The existence of Pavlovian conditioning and instrumental learning means that animals have evolved mechanisms to associate events or acts with important fitness outcomes. The association allows them to modify their behavior appropriately to promote events that cause fitness gains and to avoid events that cause fitness losses.

These two general forms of learning, active throughout an animal's life, exist because the nature is constantly in flux. In one year, heavy rains mean that a preferred food plant will be found in particular locations, but in other years, will not. A new species of predator moves into the area where an animal lives, and later moves away. Flowers that nectar-foraging animals such as honeybees visit appear, bloom, and disappear.

Even though Pavlovian conditioning and instrumental learning seem general, they are not, entirely. What an animal can learn and how quickly it can learn are influenced by its overall behavioral repertoire. Hummingbirds feed by hovering in front of a flower, inserting the beak and using the tongue to lap the few microliters of nectar that the flower offers. In the laboratory, researchers trained individual hummingbirds of a few different species to perform a simple task. The bird had to fly to a perch. When it alighted, it was presented with two artificial flowers on the other side of the room. The bird was allowed to sip from one flower, then had to return to the perch before it could drink again from either flower. The researchers trained the birds in "switch" learning, meaning that to be rewarded at the second visit to a flower, the bird had to visit the other flower, not the one that it had just visited. The researchers also trained the birds in "stay" learning, meaning that to be rewarded at the second visit to a flower, the bird had to return to the flower that it had just visited. All birds were trained at both tasks. Although these tasks seem equivalent to us, they are not equivalent to a hummingbird. In the study, the researchers found that it was relatively easy to train the birds to switch learning, and very difficult to train them to stay learning. This difference in learning rates occurred because of the natural foraging behavior of hummingbirds. In nature, a bird depletes the nectar from one flower and then moves on to another flower. An individual maximizes its rate of energy harvest specifically by not returning to a flower that it has just visited. Thus, in switch learning, the researchers were asking a bird to learn to do what it does in nature, but in stay learning were asking it to override its natural foraging preference.

Instrumental learning is what happened when Clever Hans apparently became a mathematician. His owner unconsciously trained him when to start and when to stop pawing with his forefoot. Unconscious instrumental training by owners is the basis for most of the problem behavior of pets. Instrumental learning is the basis for what animal trainers do. At a dog show, each of a dog's postures, reactions to cues, and gaits are carefully trained responses, created by the handler using instrumental learning with food rewards.

7
Moving Through Space

On a dark, moonless night in the Syrian desert, a female golden hamster emerges from her underground burrow. She pauses, with only her head above the burrow mouth, for several minutes. This pause allows her a reasonable chance of detecting predators that may be lurking nearby. Then, she trots into the dark to forage for the seeds of desert plants. Her foraging trip may last for an hour or more and carry her several hundred meters away from her burrow. When her cheek patches are filled, or when her aching mammary glands tell her that it is time to return to her hungry pups, she runs along an almost straight path back to the burrow entrance.

How does she accomplish such accurate homing? There are no visual landmarks, because the sky is cloudy and dark. There may be some chemical signposts, left by the female when she smeared her flank glands on rocks, but these would not indicate the straight-line vector back to home. There are no detectable sounds from the burrow. Despite the absence of any useful external cues, the female, at the outbound end of her erratic foraging path, somehow "knows" the direction in which the burrow lies.

Path Integration

This mysterious ability of the hamster, most vertebrate animals, and many invertebrate animals, is called *path integration*, and it is accomplished by the running addition of travel vectors. Carefully controlled experiments in the laboratory prove that hamsters are able to find home accurately when all other sources of sensory information are blocked.

In the vertebrate animals, there are three sources of information about travel vectors. First, and most important, there is the vestibular sense, provided by the vestibular organs – the saccule, utricle, and the three semicircular canals of the inner ear. Information from the vestibular organs provides very detailed information that the brain computes as acceleration in any direction. The second source is the touch sensation, created when the animal pushes against the ground, air, or water to propel itself. The third source is a reference copy of the motor commands that go to muscles. These three sources of information are integrated to provide detailed and accurate information about the direction and distance of travel segments. In the vertebrate animals, the brain area where this self-referential information is put together and used to compute where an individual is with respect to any previous start point is the hippocampus, the same area that produces cognitive maps.

Path integration is an example of a specific, reward-free learning program that operates in the background, whenever an animal moves. It is the primary sense that supports nearly all the moment-to-moment movements that animals perform. Animals always "know" where they are with respect to where they have just been. It is difficult to overemphasize the importance of this ability, even for animals that do not have a home location.

Landmarks

Invertebrate animals do not have the sophisticated vertebrate vestibular system and so must rely on the less precise mechanisms of path integration. Their more error-prone path integration gets them at least to the vicinity of home, where another class of cues, landmarks, becomes useful. Niko Tinbergen published a now classic demonstration of this phenomenon. Tinbergen worked on the bee wolf, a solitary wasp that digs a burrow in sandy soil and then hunts for a bee, which will become the paralyzed meal upon which she lays her egg in the burrow. A wasp, when leaving to hunt, flies in rough circles around the burrow for a few moments before departing to hunt. Tinbergen hypothesized that the circling flight was designed to allow the wasp to memorize landmarks close to the burrow entrance. He tested the hypothesis with his characteristic wit: he placed pine cones in a ring around a burrow while a female was inside. The female emerged, performed her orientation flight, and flew away. Tinbergen then moved the ring of pine cones about one foot away. When the female returned, she searched only within the displaced ring, one foot away from the plainly visible burrow entrance. She was able to find her burrow only after Tinbergen moved the cones back to their original location. There are thousands of solitary bee and wasp species, in which a female constructs a nest, provisions it, lays an egg, and seals the entrance. In these species, landmark learning is almost certainly the way that individuals find the nest entrance after a trip away.

Sun Compass

Karl von Frisch, who shared the Nobel Prize with Niko Tinbergen and Konrad Lorenz in 1973, discovered what he called the "dance language" of honeybees, and in doing so proved that insects can use the sun's azimuth to navigate. Honeybees are able to collect huge amounts of food because a colony sends many workers out each day to forage for nectar and pollen, and because individuals help each other to find food by communicating the location of rich food sources. They do this using the dance language discovered by von Frisch.

Von Frisch made his astounding discovery using keen powers of observation, inspired hypothesis formulation, and inventive hypothesis tests. The innovation that started the process of discovery was the observation hive. Von Frisch discovered that he could build a tall narrow box that bees would accept as a nest cavity. Further, he discovered that he could use a pane of glass as one of the wide sides, as long as he kept the glass covered most of the time. To observe the bees going about their natural behavior inside the hive, von Frisch could sit on a stool in front of the glassed side, and remove the cover. The bees would continue to act normally. When we first look into such an observation hive, we perceive what looks like chaotic activity. Bees hurry in all directions; each intent on the particular task that she is carrying out at that moment. This could be feeding larvae in brood cells, building comb, carrying out trash, and so on. What we, the observer, see is a mad scramble. However, if we have von Frisch's keen observational power, we may notice small clusters of bees. Each cluster comprises several individuals that form a circle around a single bee. The single bee is moving quickly in mirror-image half circles. She walks in a right-turning semicircle from twelve o'clock to six o'clock, then moves in a straight line back to twelve o'clock (up through the center of the clock face), rapidly swinging her abdomen from side to side, while elevating her wings slightly and vibrating them. Then she turns left and walks in another semicircle to six o'clock, does her waggle run back to twelve o'clock, turns right, and so on. Von Frisch called this performance a bee dance.

A dancing bee performs for several minutes, and, during the performance, the straight line of the waggle run always points in the same direction with respect to the vertical. The circle of bees that surrounds a dancer crowds in close, so that their antennae can touch her. The information that the observing bees are noting is first, the angle between the vector of the waggle run and the vertical, and second, the length of the waggle run. Both pieces of information are important. The dancing occurs on comb inside the hive, and thus normally in the dark. That is why the observing bees press in close. They are getting the information by feel.

Von Frisch also invented the technique of experimental feeders. He showed that he could set out a small glass dish of sugar water and the bees from his hive would come to it. While a bee was busy pumping her crop full of sugar water, von Frisch could apply a dot of paint to the top of the bee's thorax. Away from the hive, bees are not aggressive, and they tolerate being pushed around by a paintbrush.

Back at the hive, von Frisch observed the painted bees dancing. He realized that the dance must be some form of recruitment to the rich food source that the forager had just found, because increasing numbers of workers showed up at the feeder after the dancing started. Von Frisch then set out several feeders, at different directions and distances from the hive. At each feeding site, he painted the bees with a different color. Back at the observation hive, he noticed that bees of the colors corresponding to the different feeders were dancing and crucially, observed that all bees of the same color danced with the same angle of the waggle run with respect to the vertical. Von Frisch was able to accurately measure the angle between the vector of the waggle run and the vertical by laying a protractor against the glass. A weighted plumb bob was attached to the center of the protractor. Von Frisch could thus rotate the protractor until its bottom edge lined up with a waggle run, and read the angle. Bees of another color danced with another specific angle between the waggle run and the vertical. This was the key observation. Von Frisch realized that this angle must convey some information about the direction of the food source. Then, von Frisch made the final key observation. Bees painted the same color slowly changed the direction of the waggle run as the day wore on. The direction of the waggle run was like a clock hand, slowly rotating as the day passed. This last observation meant that the vector of the waggle run did not point, in some simple referential way, to the food source. Rather, it seemed that the waggle run indicated the direction of the food source with respect to some external, moving, point of reference. In nature, there is only one predictable moving point of reference: the sun.

Von Frisch formulated the hypothesis that a dancing bee indicates the direction of a food source that she has found by creating an angle between her waggle run and the vertical that, outside, matched the angle between the food source and the azimuth position of the sun. Using the techniques that he had invented, von Frisch confirmed the hypothesis. Bees use the convention that up, on the hive surface, corresponds to the position of the sun. A dance with a waggle run 30° to the left of the vertical means that the food source is currently 30° to the left of the sun. Further, von Frisch showed that the length of the waggle run corresponds to the distance between the hive and the food source. When a food source is close to the hive, within about fifty meters, the dance does not employ a waggle run. Von Frisch referred to these as round dances. The round dance simply indicates that there is food close by. For food sources beyond fifty meters, the bees begin to indicate a sun azimuth direction with the waggle dance, and the waggle run becomes longer as the distance to the food source increases. The dance is a miniaturized re-enactment of the forager's flight out to the food source. During the waggle run, the dancer even simulates flight, by elevating her wings slightly and vibrating them.

Long Distance Travel

Arctic terns nest and rear their young in the Arctic summer, then fly the length of the Atlantic to spend the northern hemisphere winter in the Antarctic summer. Recently, researchers attached advanced lightweight geo-locating devices to several birds, and were able to show the actual migration routes. The round-trip distance flown by eleven birds that the researchers tracked was 37,000 to 51,000 miles.

This kind of flight performance depends on the ability of the terns to find and use favorable winds, and also requires a fully functioning navigation system. Although arctic terns represent the extreme, they are not unique. Individuals of many bird species migrate hundreds to thousands of miles every year between nesting habitats in the northern hemisphere and wintering habitats in or close to the southern hemisphere. How do they navigate with such accuracy?

A.C. Perdeck, a colleague of Tinbergen, performed a key study that set the agenda for this field of research. Perdeck worked with starlings, a native Eurasian species. The population that Perdeck studied breeds in northern Holland, Germany, Poland, and southern Scandinavia, and in fall migrates in huge flocks, south-west along the border of the North Sea to southern Holland, Belgium, France, Britain, and Ireland. For ten years, from 1948 to 1958, Perdeck and his colleagues captured migrating starlings in the vicinity of The Hague, banded them, placed them in covered cages, flew them by airplane to Switzerland, and released them. The researchers made a prodigious effort to recapture birds. Because the chance of recovering a released bird is small, Perdeck and coworkers continued the study for ten years, until they had caught and released 11,247 birds and recovered 354.

Perdeck's data revealed that the direction that a released bird flew depended on whether the bird had made the migratory trip before in its life. Young of the current year, migrating for the first time, continued to fly to the south-west (their heading when nabbed in Holland) from Switzerland. Adults flew to the north-west, a direction appropriate to get to the migratory destination.

This result showed that the birds have two distinct navigation systems. The first, shown by the naïve young, is the ability to fly in a fixed compass direction. Juvenile starlings are programmed to fly to the south-west, and they continue to do so even when displaced. The adults, in contrast, had access to another navigation system. They seemed to know, in some sense, where they

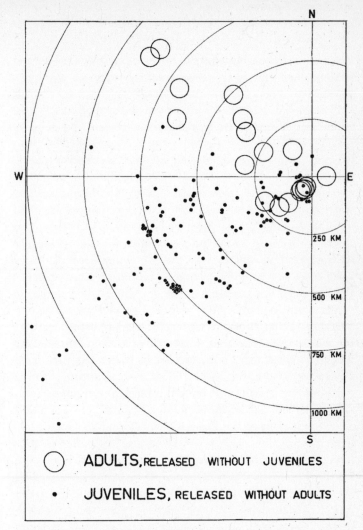

N

W   E

250 KM

500 KM

750 KM

1000 KM

S

◯   ADULTS, RELEASED WITHOUT JUVENILES

•   JUVENILES, RELEASED WITHOUT ADULTS

Fig. 8. Recoveries of fig. 5-7 coincided at one release point (recoveries in March excluded).

Figure 7.1 One of Perdeck's original figures, showing different orientation directions of adults and juveniles when released in Switzerland.

were on the globe with respect to their migratory destination, and they followed the appropriate compass heading to get there. Juvenile birds have a compass. Adults have a compass and also a map.

Subsequent work on birds followed and supported the conclusion, established by Perdeck, that there are two separate navigation systems, the compass sense, and the map sense. Of these two, the compass sense has proved far more amenable to experimental study. Early work showed that a caged bird in migratory condition shows *Zugunruhe*, a nearly constant fluttering, hopping and flapping against the cage wall in the direction that it would move if it could. When the apparent position of the sun was changed for these birds, the orientation of their *Zugunruhe* changed. Starting with this lead, researchers produced conclusive proof for a sun compass in homing pigeons. Homing pigeons are useful because they repeatedly fly home when displaced. A researcher does not need to wait for the subject to come into migratory condition and to make one of its two annual trips.

The proof came from clever experiments performed by Klaus Schmidt-Koenig of the Max Planck Institute for Behavioral Physiology, in the late 1950s. The key experiments involved clock-shifting, in which the pigeons were put on an altered light-dark cycle. The light cycle is important because the sun moves across the sky during the day; it is useful as a cue to navigate in a fixed compass direction only when its position is used in conjunction with a clock. The researchers reasoned that if the birds use the sun and an internal clock to set direction, then resetting the internal clock should cause the compass heading to change. This is exactly what the clock-shifting experiments demonstrated. For example, we clock-shift a group of pigeons six hours: they are kept indoors, and the lights come on at midnight and go off at noon. We keep the birds on this schedule for a week or so. Then, on the day of the experiment, we place the birds in covered boxes, transport them one hundred kilometers to the south, and release them, one at a time, at 6 a.m. Observers with binoculars stand on elevated towers. Each bird first flies in rough circles for a minute of two, seemingly getting its bearings, and then flies off in a straight line. The observer watches with binoculars until the bird passes out of sight over the horizon, and records the compass direction of that final sighting; this "*vanishing bearing*" is considered to be the best indicator of the direction in which the bird intends to fly. If our clock-shifted bird has an unimpaired map sense, and is motivated to fly north, in which direction should it fly? To the clock-shifted bird, 6 a.m. is subjective noon. At noon, the sun is in the south. Our clock-shifted bird, if it is using the sun's position to compute where north is, should fly directly away from the sun, or due west. On most testing days, this is exactly what the birds do.

Magnetic Compass

Sometimes, the clock-shifting experiments did not work. The birds flew straight home, giving no indication that their clocks were off. Researchers eventually realized that these results occurred only on cloudy days, when the birds could not see the sun. It seemed that the birds had some kind of backup navigation system, not dependent on the sun. Researchers proposed that the backup system must be based on the other obvious directional cue in the environment, the Earth's magnetic field.

How does one discern whether a bird's compass depends on the ability to detect the magnetic field? One removes some feathers from the bird's back and glues on a bar magnet. As a control, one can glue on a brass bar of equal weight. Somewhat more precise control can be arranged by fitting the pigeon with a harness that carries a radio-controlled electromagnet. Both sorts of experiments worked. They showed that, on cloudy days when birds could not see the sun or blue sky, altering the magnetic field around the bird's head disrupted its homeward migration. The same treatments performed on sunny days had little effect.

If birds wearing magnets on cloudy days began flying in the wrong direction and then the clouds parted, the birds immediately reversed course, flying back across the release point toward home.

Thus, pigeons and other birds have two compass systems. The first is a sun azimuth based system that works in conjunction with an internal clock. Young birds learn to calibrate the motion of the sun with their own clocks during a sensitive period in development, when they must fly with a view of the sun to accomplish the calibration. Birds appear to give priority to the sun compass, but if the sun compass is not available, then they rely on a magnetic compass.

How do birds extract useful information about compass direction from the magnetic field of the Earth? One obvious possibility is that they have the equivalent of a compass and are sensitive to polarity. The other, more arcane, possibility is that birds are able to detect magnetic inclination, the angle between magnetic field vectors and the Earth's surface. As one moves from the equator toward the pole, this angle changes, from zero at the equator where the field vectors are parallel to the Earth's surface, to 90° at the pole, where the vectors point inward to the pole. Thus, the average angular change is about 0.009° per kilometer. A detector that indicated direction toward the pole would need to be very sensitive. Nevertheless, recent work has proved that the magnetic compass of birds is an inclination detector. Experiments in which birds are exposed to manipulated magnetic fields show that a bird that intends to fly to the north will orient to the south when magnetic north polarity is unchanged but magnetic inclination is flipped. The best evidence to date indicates that a bird senses magnetic inclination with its eyes. In effect, it sees the direction toward the pole.

Stars

Many bird species that migrate away from the northern hemisphere in fall do all their long-distance flying at night. The birds rest and eat during the day, then start each leg of the migratory trip at dusk. For a little while, afterglow on the western horizon provides some compass information, but soon the sky is dark, and there are no cues to indicate the position of the sun. Also, the magnetic compass, which is light-based, is unavailable. To maintain a steady southbound heading, night flying migrants use yet another navigational sense: the ability to compute compass direction from the pattern of stars in the night sky. The first conclusive proof for star compass orientation came from the Ph.D. dissertation research of Stephen Emlen, conducted at the University of Michigan in the 1960s. Emlen worked on indigo buntings, a North American species that breeds in the eastern United States and migrates at night to Mexico and Central America.

Emlen held captive buntings until they were in migratory condition, and then monitored the directionality of *Zugunruhe* of individuals under a simulated night sky in a planetarium. In the planetarium, each bird was placed in a funnel-like apparatus that gave it a view of the simulated sky overhead, but nothing to the side. The advantage of a planetarium is that we can make the stars obey our commands. We can remove selected stars, entire constellations, or large patches of the night sky. We can change the apparent time of night, and control whether the stars continue in their normal nighttime circular path around the pole star. Finally, if we are devilishly clever, we will make all the stars rotate, not around the North Star, but around another star. Emlen did all of these things, and proved that: 1) indigo buntings in fall migratory condition try to move south as indicated by the planetarium sky, 2) indigo buntings in spring migratory condition try to move north as indicated by the planetarium sky, 3) conjunction between the time indicated by the bird's internal clock and the time indicated by the planetarium sky was unimportant, and 4) no particular stars or groups of stars were critical; any few stars somewhere in the circumpolar 35° were sufficient to allow the birds to orient correctly. Thus, it seemed that the birds knew, in some way, the spatial relationship between any pair or group of stars and the pole star. To test this hypothesis, Emlen captured newly hatched indigo buntings, brought them to the laboratory, and divided them into three experimental groups. Group One birds lived in a room where diffuse lighting went on and off to reflect normal day length. Emlen moved Group Two birds to a planetarium every night and provided them with a view of a normally rotating night sky. Emlen moved Group Three birds to a planetarium every night and provided them with an altered view of the night sky, in which the stars rotated not around the pole star, but around Betelgeuse, a bright star in the constellation Orion. Emlen tested the three groups under a normal night sky in the planetarium. Group One birds showed *Zugunruhe*, but no directional preference; their orientation was not different from random orientation. Group Two birds showed significant orientation toward south as indicated by the planetarium sky; they oriented away from the pole star. Group Three birds also showed directed, nonrandom orientation, but they oriented in the direction away from Betelgeuse. Thus, Emlen proved that indigo buntings must learn how star patterns indicate direction, and that they do so by observing, when they are young, the axis around which all the nighttime stars appear to rotate.

Magnetic Map?

In a typical homing experiment, in which a pigeon is transported in a covered cage in an automobile and released perhaps a hundred kilometers from home, the bird shows little hesitation about where to go. The bird flies in rough circles for a minute or less and then heads away along a straight-line path. The direction that the bird chooses may be incorrect (because the bird was clock-shifted), but there is little apparent indecision in choosing that direction. Experiments in which the compass is disrupted by clock shifting could only work if the subjects had unimpaired map information. If a bird is clock shifted six hours early and released south of home at dawn, it flies west, away from the sun, because it intends to fly north. The clock-shifted orientation shows quite clearly that the compass sense is independent of the map sense.

How does the map sense work? At present, the short answer is that we don't know, but that the sense seems to be based on detection of some aspects of the Earth's magnetic field. Researchers have proposed that an effective map sense must be based on a bi-coordinate system that is analogous in some way to the human bi-coordinate system of latitude and longitude. They have pointed out that a bi-coordinate system of sorts is available to animals that can detect magnetic inclination and magnetic field intensity. When these two aspects of the Earth's magnetic field are plotted, they form a rough two-dimensional grid. Across large areas of the Earth's surface, location can be represented by the local readings of magnetic field inclination and intensity.

Some of the best evidence for the magnetic map hypothesis comes from an innovative study done by Kenneth Lohmann and his colleagues at the University of North Carolina. The Lohmann group has worked on the orientation of sea turtles, and especially loggerhead turtles, for many years. After it hatches in the warm sand of a Florida beach, a baby loggerhead turtle hurries down to the sea, and begins to swim east. If it is not picked off the beach or out of the shallows by the gulls that hover overhead, it continues to swim east until it meets the Gulf Stream, the northeast-bound part of a vast circular current, the North Atlantic Gyre. For the next few years, the turtle feeds and grows within the Gyre. The turtle needs to remain within the Gyre's latitudinal limits. If it strays north of the Gyre, it will die in the rapidly cooling water. A turtle that strays to the south risks being carried into the South Atlantic current system, which would take it far from the habitat to which it is adapted, and from the beaches to which it will return to lay eggs.

The Lohmann group tested the hypothesis that young turtles could detect magnetic field inclinations and intensities that indicated that they were in danger of leaving the North Atlantic Gyre. They outfitted hatchlings that had never been in the ocean with harnesses that held the turtles in place in a circular tank in the laboratory, but allowed them to orient and swim toward any direction. Next, the researchers used magnetic coils that surrounded the tank to create field inclinations and intensities that were characteristic of one of three, far-apart, locations at the edge of the North Atlantic Gyre. The results showed that the young turtles in effect knew where they were. In response to magnetic fields characteristic of the locations, the turtles always swam in directions that would carry them toward the center of the Gyre. Thus, loggerhead turtles, when they enter the ocean for the first time, are equipped with the ability to detect magnetic field inclination and intensity, and to use this information to swim in a direction that will keep them within an appropriate habitat.

More recently, several groups of researchers have manipulated the magnetic fields around birds captured during migration. Birds given magnetic information indicating that they are at the start of a long flight across water or desert eat more, and put on more fat than control birds that are given magnetic information indicating their true location. In one of the most amusing of these studies, researchers captured juvenile wheatears in Sweden, and subjected the experimental group to magnetic information indicating that they were in the Atlantic Ocean, south of Greenland and, over the next few days, were moving due south. The control birds were given magnetic information that indicated that they were on a normal migratory path to Africa. The experimental birds ate more, and put on fat. The magnetic map information told them that they were in the middle of a long ocean flight, and they ate accordingly, apparently not too confused by the wooden shed they were living in, the mealworms provided, and the absence of any kind of ocean view.

8

Genetics

The blackcap (*Sylvia atricapilla*) is a common, widespread Eurasian warbler. In fall, individuals that spent the summer in Europe migrate to North Africa. A bird migrating to Morocco from France obviously has a much shorter trip than a bird migrating from Finland. Researchers captured nestlings from four populations: Finland, Germany, France, and the Canary Islands. They raised the birds in captivity in Germany under uniform conditions. In fall, the researchers monitored the duration of *Zugunruhe* of the birds, and found that individuals from the Finnish population had the longest durations, followed by the German, the French, and finally by the Canary Island birds. These results suggested that the birds of these four populations differed genetically in the programmed duration of migration. To provide further proof of this hypothesis, the researchers mated German birds to Canary Island birds, and recorded the duration of *Zugunruhe* of the hybrid offspring. The hybrids, with a German father and a Canary Island mother or vice versa, had *Zugunruhe* durations that were midway between the parental values.

These results are typical of the results one gets when one creates hybrids from two distinct parental lines that show genetically based differences in behavior. The behavior of the hybrid offspring is usually an average of the values of the mother and the father. From these findings, we can conclude two things about the genetic control of behavior. First, genetic differences among individuals contribute significantly to the total variation in behavior among individuals. Second, most measurable aspects of behavior, such as the duration of *Zugunruhe*, are controlled by many genes.

If a behavioral phenotype is determined by only one or two genes, then hybrid offspring would not show continuous variation: they would fall into discrete bins. For example, suppose that the duration of *Zugunruhe* is controlled by variation in a single gene. Each individual bird, like most animals, is diploid: each of its cells contains two copies of each chromosome and thus two copies of each gene. A chromosome is a single long molecule of DNA comprising hundreds to thousands of genes. Genetic variation exists because a gene copy may come in several different forms. In other words, the sequence of nucleotides that constitute the gene can vary somewhat. This variation is created by errors in DNA copying when eggs or sperm are made. These copying errors are called mutations. For the purposes of illustration, let's assume that the *Zugunruhe* gene comes in two forms, which we will designate as A and a. Because each individual has two copies of each gene, an individual's genotype may be AA, Aa, or aa. Let's assume that a bird with a genotype of AA has a *Zugunruhe* duration of eight days, a bird with an Aa genotype a duration of six days, and a bird with an aa genotype a duration of four days. When a male bird of genotype Aa mates with a female of genotype Aa, we can keep track of possible genotypes of the baby birds with a mating table (Table 8.1) that shows the possible sperm and egg combinations. The four possible offspring genotypes are shown in the cell entries.

Table 8.1 Mating table, showing the offspring genotypes that are produced by mating between a father with genotype Aa and mother with genotype Aa.

| | | Egg Types | |
		A	a
Sperm	A	AA	Aa
Types	a	Aa	aa

One-fourth of the offspring will be genotype AA, with *Zugunruhe* durations of eight days, one-half will be genotype Aa, with *Zugunruhe* durations of six days, and the remaining one-fourth will be genotype aa, with *Zugunruhe* durations of four days. There are no intermediates, with *Zugunruhe* durations of seven or five days. It was by following pedigrees and observing proportions like these in garden plants, that Gregor Mendel deduced the existence of genes.

Phenotypes of individuals fall into discrete bins when a single gene controls the trait being measured. What happens if two genes are in control? If the second gene can also take one of two alternate forms (the alternate forms of genes are called alleles), which we will designate as B and b, there are nine possible offspring genotypes, as shown in Table 8.2. Table 8.2 also shows hypothetical values for *Zugunruhe* duration of each genotype.

We figure out the proportions in which these genotypes will appear by constructing a mating table, as we did to get genotype frequencies for a single gene. If we count the genotypes in Table 8.3, we get the following result:

Table 8.2 When two genes, each with two alleles, control a trait, there are nine possible genotypes. Values below each genotype show hypothetical *Zugunruhe* durations.

| | | B Genotype | | |
		BB	Bb	bb
A Genotype	AA	AABB 10	AABb 9	AAbb 8
	Aa	AaBB 8	AaBb 7	Aabb 6
	aa	aaBB 6	aaBb 5	aabb 4

Table 8.3 A mating table, showing the offspring genotypes that are produced when two genes, each with two alleles, control a trait.

		Egg Types			
		AB	Ab	aB	ab
Sperm Types	AB	AABB	AABb	AaBB	AaBb
	Ab	AABb	AAbb	AaBb	Aabb
	aB	AaBB	AaBb	aaBB	aaBb
	ab	AaBb	Aabb	aaBb	aabb

We can also view the expected spread of *Zugunruhe* durations with a frequency histogram, a plot with trait values on the x-axis and the number of observations of that value on the y-axis (Figure 8.1).

Table 8.4 When two genes, each with two alleles, control the duration of *Zugunruhe* genotypes appear in numbers as shown. *Zugunruhe* durations are transcribed from Table 8.2.

Genotype	Number	Zugunruhe *Duration*
AABB	1	10
AABb	2	9
AAbb	1	8
AaBB	2	8
AaBb	4	7
Aabb	2	6
aaBB	1	6
aaBb	2	5
aabb	1	4

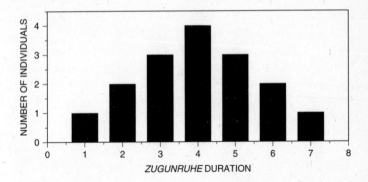

Figure 8.1 Frequency distribution of *Zugunruhe* durations when two genes, each with two alleles, control the trait.

If we add a third gene, also with two alleles, there are twenty-seven possible genotypes. If ten genes, each with two alleles, control the trait, we have 59,049 possible genotypes. There are so many bins that all the spaces between bins that exist in Figure 8.1 become filled in, and we can replace the discrete frequency distribution with a continuous one: the familiar bell curve, or Gaussian distribution, that describes the array of values of many things in nature. Most behavioral traits, like *Zugunruhe* duration, when measured in many individuals, have this kind of frequency distribution. This tells us that multiple genes control most behavioral traits.

This should not be too surprising, because behavior is the result of action potentials that travel along motor nerves to muscles. The brains that make the decisions about which muscles to contract from moment to moment are complex organs. Brains and other organs are built by the developmental process in which a single fertilized egg divides, and its daughter cells divide, and those granddaughter cells divide, and so on until there are enough to make up the blocks of primordial cells that will become guts, muscles, bone, skin, nervous system, and so on. These incipient tissues participate in a developmental program to take on species-specific structure and function. The developmental program is complex, and is regulated by the closely timed and coordinated expression of hundreds of genes. Thus, it is generally unlikely that a connection between a single gene and a behavioral outcome could be found. Genes do not cause behavior. Genes build brains, and brains cause behavior.

Even though most behavioral acts emerge from the expression of many genes, we can still study how these traits are inherited, and form estimates of the rate at which behavioral change will occur when there are new selection pressures that are caused by changes in the natural environment. Traits that are controlled by many genes, and which therefore vary continuously (no discrete bins) are called quantitative traits, and the study of the genetics of such traits is called quantitative genetics. Quantitative genetics relies heavily on two statistics that one can calculate from continuously varying data: the mean and the variance.

The mean, or average, is familiar to many people. We sum all the individual values in our sample and divide that sum by the number of individuals that we sampled. For example, to calculate the mean from the data shown in Table 8.4 and Figure 8.1 we take $(1 \times 4) + (2 \times 5) + (3 \times 6) + (4 \times 7) + (3 \times 8) + (2 \times 9) + (1 \times 10) = 112$ and divide this result by the number of individuals (16) to get the mean of 7. The mean calculated this way is called the arithmetic mean, and is often an accurate descriptor of the central tendency in a group of data, as it is when we look at Figure 8.1. Calculation of an estimate of the variance is only a bit more complicated. From a discrete sample, the variance is defined as the average squared difference from the mean. To calculate the variance for the data shown in Table 8.4 and Figure 8.1, we take $1 \times (4-7)^2 + 2 \times (5-7)^2 + 3 \times (6-7)^2 + 4 \times (7-7)^2 + 3 \times (8-7)^2 + 2 \times (9-7)^2 + 1 \times (10-7)^2 = 40$, and divide by the number of samples (16) to get 2.5. The data shown in Table 8.4 and Figure 8.1 have a mean of 7 and a variance of 2.5. Two different samples could have the same mean but different variances. The variance describes the degree of spread around the mean. A large variance means that there are many values far from the mean and a small variance means that most values are clustered close to the mean.

When we measure a behavioral variable in a sample of individuals, the variance that we calculate is called total, or *phenotypic* variance. The quantitative genetic approach considers phenotypic variance to be a sum that is composed of two main parts: environmental variance and genetic variance. Suppose, for example, we measure IQ in a thousand humans. The distribution of IQ values will be Gaussian (a bell curve). A small number of individuals will have very low IQ scores and an equally small number will have very high IQ scores. Most individuals will cluster around the mean IQ score. What causes the variation in IQ score? On the one hand, we could assert that all the variation must be due to the environment, including effects such as the mother's nutrition during gestation, postnatal nutrition, exposure to toxins, illness, parental involvement in education, and so on. Alternatively, we could assert that all the variation is due to variation in individual genotype. For IQ, as for nearly all behavioral traits, the answer is that both environment and genotype contribute to the total variation, which we write as $V_P = V_E + V_G$: phenotypic variance equals variance due to environmental effects plus variance due to genotype. Furthermore, geneticists know that V_G can be represented as the sum of three components. These are additive effects, V_A, dominance effects, V_D, and interactive effects, V_I. So, we write $V_G = V_A + V_D + V_I$.

To understand these separate components of genetic variance, look again at Table 8.2. This shows that individuals with different genotypes have different *Zugunruhe* durations. The data in Table 8.2 were made up in a special way to show all additive genetic variance. Notice that, anywhere in the table, the effect of replacing an a allele with an A allele is an increase of two days. The effect of replacing a b allele with a B allele is always an increase of one. In Table 8.2, all the effects on phenotype are additive genetic: the effects of individual alleles simply add up. If genetic dominance existed, then, for example, the genotype AaBB (row 2, column 1) would have a value of ten, just like AABB, because only one copy of the A allele is needed to get a full effect. Interactive effects, also called epistasis effects, mean that the effect of one allele depends on the genotype for another gene. For example, note that all the values in row 2 (Aa genotypes) increase by two in row 1 (AA genotypes); the effect of replacing a with A is always an increase of two. To illustrate an interactive effect, change the value in row 1 column 3 to six. If we compare row 2, column 3 to row 1, column 3, the effect of replacing an a allele with an A allele is no longer two, but zero. If the B genotype is BB, or Bb, the effect of replacing a with A is still two, but if the B genotype is bb, this A effect disappears. The effect of the A genotype was dependent on the B genotype. That is an interactive effect.

Of these three genetic effects, additive, dominance, and interactive, quantitative geneticists tend to be most interested in additive effects. That is because additive effects are responsible for the resemblance between parents and offspring. Let's go back to Table 8.2, which, remember, illustrates all additive effects. Let's create a few matings between parents with different *Zugunruhe* values and compare the mid-parent value (average of the two parents) to the offspring value (Table 8.5).

If we plot the offspring values against mid-parent values, we achieve the plot in Figure 8.2

The line in Figure 8.2 is at 45° to the horizontal. It has a slope of 1, meaning that, to stay on the line, for every increase along the x axis, we must make the same increase along the y axis. The slope of this line is equal to an important value in quantitative genetics called *heritability*, and designated as h^2. In reality, when we measure mid-parent and offspring values, we don't get a set of

Table 8.5 From Table 8.2, we take values for parental genotypes, generate the offspring genotypes and find offspring values from the same table.

MALE PARENT		FEMALE PARENT		MID-PARENT *ZUGUNRUHE*	OFFSPRING	
Genotype	Zugunruhe	Genotype	Zugunruhe		Genotype	Zugunruhe
aaBb	5	aabb	4	(5+4)/2 = 4.5	aaBb	5
					aabb	4
						mean = 4.5
AAbb	8	aabb	4	(8+4)/2 = 6	Aabb	6
AABB	10	aabb	4	(10+4)/2 = 7	AaBb	7
AABB	10	aaBB	6	(10+6)/2 = 8	AaBB	8

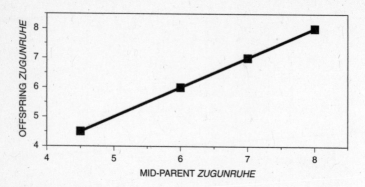

Figure 8.2 Plot of mid-parent values against offspring values of *Zugunruhe*. The data were taken from Table 8.5.

points that fall neatly on the line as in Figure 8.2. Instead, we get scatter, as shown in Figure 8.3

The line drawn through the cloud of points is derived from a statistical procedure which finds the straight line that results in the minimum sum of squared differences between each point and the line. This is called a least-squares regression line, and the slope

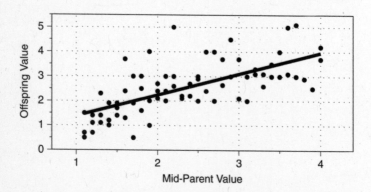

Figure 8.3 A typical plot of mid-parent values on offspring values for a behavioral trait.

of this line is equal to the heritability. In Figure 8.3, this value is 0.85. Another way to express heritability is as a ratio, defined as V_A/V_P. In other words, the heritability of a trait is the proportion of total phenotypic variance that is attributable to the additive effects of alleles. Because Table 8.2 was designed to show only additive genetic effects, the mid-parent-offspring plot from the data show a slope of 1. The heritability is 1, meaning that 100% of the variance in *Zugunruhe* duration is attributable to additive effects of alleles. In the relationship shown in Figure 8.3, the slope of the least-squares regression line is 0.85, so 85% of the phenotypic variance is attributable to additive effects of alleles, and the remaining 15% is attributable to some combination of environmental effects and non–additive genetic effects (V_D and V_I).

Heritabilities have been measured for hundreds of behavioral traits, including locomotion, nesting, social dominance, aggressiveness, vigilance, grouping tendency, exploratory behavior, fearfulness, sensitivity to pain, maternal solicitude, willingness to help parents raise new offspring rather than breed for oneself, social play, wheel running, and mating speed, to list a few examples. The heritabilities range from about 0.10 to 0.75, with most falling in a range from 0.3 to 0.5.

Because most measured behavioral traits have significant heritabilities, it should be possible to modify behavior by selective breeding, and it is. There are many examples of such modification. First, as Charles Darwin pointed out in the first chapter of *The Origin of Species*, many differences in behavior are apparent among the domestic varieties of animals, such as pigeons, chickens, cattle, and sheep. Darwin paid special attention to the huge variety among domestic pigeons, all of which are descendants of the wild rock dove. Along with a riot of sizes, shapes, colors, and plumage, the breeds of domestic pigeons also show many differences in behavior. Pigeon breeds vary in vocalizations, postures, walking gait, flying maneuvers, flight speed, and homing ability, and even the ability to fly (some breeds cannot). Humans created these differences by controlling the breeding of their pigeons, allowing individuals with desirable (as defined by the human owners) behavioral traits to reproduce, and culling animals with undesirable behavior.

Another example of change under domestication occurs in the domestic sheep of North America. In western North America, domestic sheep are unusually vulnerable to coyotes. When a sheep is grasped about the neck by a coyote's jaws, it tends to go limp, as if it is saying to the coyote, "I give up: go ahead and eat me." This is odd, because a domestic sheep weighs a lot more than a coyote. If an individual struggled vigorously, it might be able to fend off a coyote attack. However, individuals don't struggle, and so the coyote enjoys an easy meal. How can this be?

The answer is that selective breeding inherent in the domestication process modified the behavior of domestic sheep. Since the start of large-scale sheep shearing, humans have selected for an immobility response when the shearer grasps the sheep and plants it on its rump between his legs. Sheep that struggled and would not hold still for shearing were not allowed to breed. In consequence, modern domestic sheep are relatively easy to shear. However, the immobility response when grasped that makes them easy to shear also makes them fail to struggle when they are grasped by a coyote.

Making Nice Foxes

In a few cases, the behavioral outcomes of the domestication process have been studied more systematically. One of the most famous of these studies is the silver fox experiment, begun in 1959 by the Russian geneticist, Dmitry Belyaev. Belyaev wanted to replicate the process by which a wild species becomes domestic. He started a selective breeding experiment on foxes taken from an Estonian fox farm. The foxes at the start of Belyaev's experiment were hostile and fearful toward humans. Belyaev's intent was to select on a single behavioral trait: tameness toward humans.

Each generation, researchers tested juveniles for tameness, starting at age one month, and then at monthly intervals. At seven to eight months of age, when individuals were sexually mature, the researchers assigned each to one of three tameness classes. A fox placed into class III flees humans, and when cornered in the cage, crouches, snarls, and snaps. A fox placed into class II can be touched, but it only tolerates the touch, and shows no attraction to humans. A fox placed into class I approaches the experimenter, wags its tail, and whines. In each generation, only the tamest individuals, generally less than 10% of those tested, were allowed to breed. Belyaev died in 1985, but some of his colleagues continued the selection experiment. After more than thirty generations of selection and 50,000 tested individuals, the foxes have been transformed. Most individuals will jump into the researcher's arms and lick her face. They are affectionate, eager to please, and compete with each other for human attention.

In a few decades, the Belyaev experiment has re-created the process, which likely took place over thousands of years, through which the wolf was transformed into the domestic dog. The Belyaev experiment achieved quicker results because the selection on a behavioral trait was rigorous, based on repeated testing of individuals, and quite severe, allowing less than 10% of individuals each generation, only the most tame of those tested, to breed. In the domestication process that led to the domestic dog, selection was much more intermittent and haphazard, and was certainly not based on the repeated, formal testing of individuals. Nevertheless, the élite foxes of the Belyaev experiment today are strikingly like domestic dogs in their affection for humans, their ability to follow human hand signals, their tendency for white splotches on the coat, floppy ears, curled tails, a shortened muzzle, and in their delayed postnatal development, which provides for a longer period during which bonding to humans can occur. These results provide another reminder that it is misleading to think about "genes for behavior." The genetic changes that resulted from the Belyaev selection study were those that affected the timing of development, hormones that regulate the stress response, and others not yet identified. Belyaev selected on a single behavioral trait, tameness, and in doing so caused comprehensive developmental and genetic changes that affected far more than the specific behavior that was the target of selection.

ADHD Mice

Another research project, involving selection on the behavior of captive animals, one that rivals the silver fox experiment, is the mouse wheel running study. This work, performed by Ted Garland and many collaborators, has extended for almost twice as many generations as the fox study, and been even more thorough in the examination of the correlated changes that accompany selection on a behavioral trait. The primary logistical issue in any attempt to practice selection on a behavioral trait is that the trait must be measured. Measurement becomes a serious consideration when, in each generation, you want to give a score to hundreds of individuals so that you can identify the top end of the distribution: the few individuals with extreme scores that will become the breeders for the next generation. The silver fox study was heroic, because the researchers invested the countless hours needed to assign a tameness score to each fox. In the wheel running study, measurement was automated. Garland and colleagues tested mice at six to eight weeks of age. A mouse was placed in a standard plastic laboratory cage that had a hole cut in the side. A short tube inserted into the hole gave the mouse an access tunnel into the activity wheel. An activity wheel is a squat cylinder, about 36 cm in diameter, with wire mesh sides and clear plastic ends.

An axle runs through the center of the ends, and the cylinder is mounted, like a car tire, so that it can spin freely.

A mouse placed into one of these cages will, during its activity period, voluntarily enter the wheel and run. When the mouse runs on the wire mesh, the wheel rotates and a photocell attached to a computer provides a way to count the number of revolutions of the wheel. Each generation, the Garland group measured the wheel running activity of about six hundred mice. Each mouse spent six consecutive days in a wheel cage, and the researchers took as its final score the average number of daily wheel revolutions on days five and six. At any time during measurement, the experimental rooms had two hundred wheels running.

The main reason why so many mice had to run was that the Garland group worked with replicated lines. A line is a population of mice maintained in the laboratory as an isolated breeding unit. Each line is a selection experiment and thus an independent replication of the experiment. The Garland group elected to create eight lines. In each of four experimental lines, individual mice were chosen for breeding based on their wheel running scores. Only the highest-scoring individuals in each family were allowed to breed. The other four lines were controls. In each control line, individuals were selected randomly for breeding, disregarding their wheel running scores. The purpose of the control lines is to be sure that any change observed in the experimental lines is due to selection, and not due to some unknown change in laboratory conditions. Control lines also provide a reference point against which changes in the selected lines can be measured.

Results for the Garland experiment, through sixty generations, are shown in Figure 8.4. At the start of the experiment, all mice ran at about 5,000 revolutions over the two-day test period. Control lines continued to run at about this level, but the experimental lines showed a rapid response to selection. By about generation sixteen, mice in the experimental lines ran more than

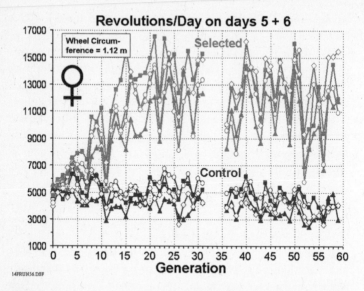

Figure 8.4 The results of the Garland wheel running selection experiment show the mean wheel running of female mice on days five and six of a six-day exposure to wheels. The experiment comprised four replicate lines of mice that were selectively bred for high voluntary wheel-running behavior and four unselected control lines. The gap in the lines at generation 32 reflects a four-generation interval when Garland and his mice moved from the University of Wisconsin-Madison to the University of California, Riverside. Figure courtesy of Theodore Garland, Jr.

twice the original level, and this elevated running persisted over the next forty-four generations.

The Garland experiment is unique in its longevity, but is even more so in the thoroughness with which the team explored the consequences of selection on a behavioral trait. Working with many collaborators, Garland examined how, exactly, the selected mice created more revolutions per day, how they were motivated to do so, whether other behavioral traits had been affected, and whether aspects of anatomy and physiology had changed.

The selected mice created more wheel revolutions per day because, compared to mice of the control lines, they ran more quickly, over shorter intervals, and with shorter breaks between intervals. In their home cages, they tended to move more, and when their running wheels were locked, they showed behavioral and brain changes similar to those found in addictive withdrawal. Garland and colleagues concluded that their selection for elevated wheel running had created a mouse version of human Attention Deficit Hyperactivity Disorder (ADHD). Although some aspects of muscle anatomy and physiology changed, most other aspects of behavior remained unchanged.

The Garland study was powerful proof of the central premise of animal behavior that evolution will act on species to create individuals that act as if they are constantly asking themselves the question, "what should I be doing at this moment to maximize my lifetime fitness?" Suppose, for example, the environment of a wild mouse species changed so that food became more widely distributed. Individuals that were more active, more likely to move, would be the most successful foragers. The Garland study shows that natural selection could cause an increase in activity without changing other importantly tuned behavioral priorities and responses.

Our current knowledge of behavioral genetics also provides another perspective on the nature–nurture debate. Is a particular behavioral predisposition built in, or is it learned? Behavioral genetics, complementing studies of behavioral development, shows that the answer is both. Most measured heritabilities of behavioral traits lie between 0.3 and 0.5, meaning that about one-third to half of the variation among individuals is due to the additive effects of alleles. The remaining half to two-thirds of the variation is due to non-additive genetic effects and to environmental effects. Nature–nurture is a false, misleading, obfuscating, dichotomy. Genes direct the assembly of brains, and brain assembly, before and after birth, is often modified by

9

Living in Groups

Some species, such as humans, are group living, but many are not. In one of our close relatives, the orangutan, individuals are alone in the forest for most of their lives. A survey of all the animals that have brains (worms, mollusks, arthropods, and vertebrates) reveals that in most species, individuals do not live in groups. Group living is the exception, not the rule, because animals in groups must compete with other group members for resources such as food, nesting sites, resting sites, and mates. Animals in groups are more likely than solitary animals to be infected by diseases or parasites carried by other group members. Thus, in most species, the best option for individuals is to live alone. Why, then, does group living exist at all? What benefit could outweigh the costs?

Why Live in a Group?

The famous theoretical biologist, William D. Hamilton, provided the answer in a charming 1971 article: *Geometry for the Selfish Herd*. The essence of Hamilton's logic was that an individual, simply by joining a group, dilutes its own likelihood of being killed by a predator. Thus, insofar as individuals of a species have a mortality risk that is due to predation, they should form groups.

Hamilton's explanation for the evolution of grouping tendency makes a simple, powerful prediction. Groups should be common among prey species, and groups should be rare in predatory species. The severity of individual risk of death due to predation should be the main driver of group formation. The main prediction of Hamilton's selfish herd hypothesis is amply verified.

Hamilton deliberately used the word "herd," a word that usually denotes a group of hoofed mammals. Nearly all the hoofed mammals are pursued by several species of predators, and nearly all form herds. Often, herds contain hundreds, thousands, or sometimes hundreds of thousands of individuals. Each individual, by joining the huge group, vastly dilutes its own risk of death by predation. In the open ocean, huge herds of fish are always formed of prey species. One never sees herds of hundreds or thousands of lions or sharks. The vast difference in the grouping tendencies of prey versus predator species is a consequence of individuals behaving selfishly. For most predators, the selfish optimum is to be solitary, to avoid sharing food with others. For most prey species, the selfish optimum is to crowd in close to others, even when this means giving up some food and being exposed to more disease. Better to be hungry and sick, with at least some chance of reproducing, than to be a corpse fed upon by a leopard, with no chance of reproducing. Herbert Prins, a biologist who studied African buffalo for many years, provided a clear illustration of this trade-off.

Social Dominance

Even though buffalo are large and powerful and pugnacious, lions can kill them. Thus, they form herds. A foraging group of several hundred individuals assumes the shape of a narrow parabola, moving in the direction of the apex. Leading animals are at the apex and trailing animals are at the rear, open side of the parabola. Prins made visual assessments of the body condition of buffalo in these foraging groups. Individual buffalo at the leading edge of the foraging groups were in the best condition, and those at the rear were in the worst condition. At the front, individuals find fresh grass. At the rear, individuals find not-so-fresh grass that has been picked over and trampled by the preceding two hundred buffalo. Differences in condition would thus emerge if individual position in a foraging group is consistent over time. In other words, some individual buffalo are always at the front of a foraging group, getting the first pick at a patch of grass as a group approaches it. Other individuals are always consigned to the rear, where they try to find something to eat in the trampled, dung-spattered leavings of the herd. Trailing animals are not able to move up in a foraging group because the individuals in front of them will not yield position. Each individual is trying to maximize its own lifetime reproduction and therefore is unwilling to share. A buffalo herd is indeed a selfish herd, as are the herds of most animal species. Some buffalo have high social status and they are able to claim a place at the front edge of a foraging herd. Others have low social status and they are not able to follow their foraging preferences because others prevent them from doing so.

The individual differences in status among buffalo are the result of a phenomenon known as *social dominance*. Social dominance is the primary organizing force in animal societies, and it is nearly ubiquitous in the species that form groups. The Norwegian biologist, Thorlief Schjelderup-Ebbe, first described it in his 1921 Ph.D. dissertation on the behavior of domestic chickens. Schjelderup-Ebbe noted that if ten or so hens that were strangers to each other were placed together in a pen, a flurry of short-lived attacks and retreats would ensue. Hen A, with her head held high, quickly steps toward hen B and pecks at her. The beak may or may not make actual physical contact. Hen B turns and steps away. Typically, what happens next is that hen A now turns and pecks at hen C, or hen C, having seen hen B lose, approaches and pecks at B. After a day or so, these overt attacks subside, as the hens settle into an established peck order (Schjelderup-Ebbe's term). In an established peck order, assertions of dominance and signals of subordinate status continue, but they become more fleeting and subtle.

In a group of ten or fewer hens, the peck order will be linear, meaning that one hen at the top dominates all other hens, there

Table 9.1 Win-loss matrix, showing counts of dominance encounters between all pairs in a five-member group.

		LOSER				
		A	B	C	D	E
W I N N E R	A		5	7	4	6
	B			8	6	4
	C				5	7
	D					3
	E					

is a hen second in rank that dominates all other hens except number one, and so on until we arrive at the bottom ranked hen, that dominates no other hens and is the target of all. The data from one hour of observation of wins and losses in these brief, formal-looking encounters for a group of five hens would look something like Table 9.1.

The dominance hierarchy shown in Table 9.1 is perfectly linear, but not all dominance hierarchies are. For example, data representing encounters of five individuals in a different group might look like Table 9.2.

The dominance hierarchy in Table 9.2 is not perfectly linear. It contains one set of relationships that is called a *circular element*. The circular element here is that B is dominant to C, and C is dominant to D, but D is dominant to B. In a win–loss matrix, circular elements are revealed by entries that appear below the principal diagonal (designated by the shaded cells). Generally, as group size increases, the number of circular elements increases.

In essentially all but the most transitory animal groups, social dominance is apparent. As for buffalo, social dominance creates inequalities. Dominant individuals have priority of access to

Table 9.2 Win-loss matrix for another group, in which the dominance hierarchy is not perfectly linear.

		LOSER				
		A	B	C	D	E
W	A		6	7	5	8
I N	B			7	0	10
N E	C				8	6
R	D		8			6
	E					

resources and subordinate individuals do not. Specialized, subtle movements that signal domination or subordination exist, and these signals regularly pass between individuals. As a result, social groups usually appear to function with apparent ease, with little overt tension to ruffle the smooth facade of harmony. However, social dominance, and the signals that support it, do not exist to create a group benefit. They exist because individuals are following their own selfish interests, and those interests include minimizing the time and energy wasted on disputes. Subordinate individuals accept their daily rounds of humiliation and their diminished access to resources because the alternative would be to live apart from the risk-diluting safety of the group.

What determines the dominance status that an individual acquires? We do not have a complete answer to this question, but we do know of two main effects. The first is obvious: size. In a social group, dominance status generally reflects relative body mass. Larger individuals become dominant. Although they may never fight with others, their larger size predicts that they would win if a fight occurred. Animals usually are quite skilled at assessing the fighting ability of a rival by observation alone, and this skill results in the acknowledgment of social rank. The dominance ranking of two individuals is essentially an agreed-upon concession of what the outcome of a fight would be. Dominance rank is not always completely accurate, and an individual that challenges its rank by fighting may sometimes rise in rank. The existence of dominance hierarchies illustrates the fact that animals in nature are very reluctant to fight. Fighting is rare because in most disputes, the fitness benefit of winning does not outweigh the fitness cost, which is expressed as the risk of injury or death.

The other main effect of dominance status is prior experience. An animal that starts life with a series of winning encounters tends to become dominant. An animal that starts life with a series of losing encounters tends to become subordinate. In essence, animals are programmed to adjust their assessment of the fighting ability of a rival based on their own personal history. This mechanism, which has been demonstrated in many species, is the primary basis for the stability of dominance hierarchies. Individuals start down one track or another, and the self-fulfilling nature of the prior experience adjustment mechanism keeps them on that track.

This effect was first shown in a laboratory experiment on inbred lines of house mice. Inbred lines are produced by many successive generations of brother–sister matings. The result is a stock of mice that are all genetically identical. Inbred lines are useful, because any differences that are observed among individuals must be due to environmental effects. Ginsburg and Allee gave one group of mice a series of losing encounters by pairing them time after time in a neutral arena with an older, heavier mouse or a mouse of a more aggressive strain. They gave another group a series of winning encounters by pairing them with a series of younger, lighter opponents or a mouse of a less aggressive strain. Then they paired individuals from the two groups. In each pairing, the two mice were identical in age, weight, and genotype; they differed only in the nature of their prior experience.

Prior experience predicted perfectly which mouse would win: winners always won, and losers always lost.

Another kind of prior experience effect occurs in many species of monkeys and apes. Most primates (except orangutans) live in social groups comprising several mother families (females and their female descendants) and several unrelated adult males. When they reach sexual maturity, males born into a group are usually evicted by older resident males and forced to find a new group to join. Dominance is well developed in primates and it is often reinforced by facial expressions or slight shifts in posture. Dominance is also a characteristic of mother families. A female typically rises to a social rank that is just below that of her mother, and so one entire mother family can be dominant to another. A young female reaches her final rank because mothers intervene in dominance disputes between juveniles. A mother can only intervene on her daughter's behalf if she herself is dominant to the mother of the other juvenile. Thus, a young female becomes dominant to all female peers whose mothers are subordinate to her own mother.

Territory

Social dominance is selfish behavior designed to ensure priority of access to potentially contested resources at any future time. A similar kind of behavior, that also works to lay advance claim on resources, is called *territoriality*. The essence of territorial behavior is that an animal actively defends the boundaries of a defended area. Defense takes the forms of advertisement that the space is occupied (birdsong, for example), patrolling to monitor territory boundaries, and aggressive eviction of intruders. Territorial behavior thus has a cost, expressed in terms of the time and energy that are devoted to defense, that ultimately can be reflected as a potential decrease in fitness. Counterbalancing the cost must be a greater fitness benefit. The benefit comes about by the exclusive use of the food, refuge, nursery sites, or breeding opportunity that the defense of the territory provides.

Territoriality has been described in hundreds of animal species, including limpets, crickets, butterflies, flies, ants, wasps, bees, spiders, snapping shrimp, frogs, salamanders, turtles, alligators, lizards, fish, birds, and mammals. The resources claimed by territorial defense vary from mating opportunity alone to all-purpose areas where mating, feeding brood rearing, and sheltering occur. Similarly, territory sizes vary greatly, from tiny areas equal in diameter to about four to five of the owner's body lengths, to huge areas, equal in diameter to thousands of the owner's body length.

Limpets: Bumper Cars in a Washing Machine

Limpets are snails that live on rocks at the ocean tideline. They use a muscular foot to create the suction that keeps them in place on wave-pounded vertical slabs of rock. A limpet crawls slowly across the rock and uses its radula, a tongue-like scraping organ, to harvest the algae that carpet the wet rocks. Limpets are territorial. Each individual defends a patch of rock and by keeping others out, ensures that the algae are harvested at a rate appropriate to maintain a steady supply. Territory owners shove intruders and may dislodge them from the rock, casting them into the surf below, to almost certain death.

Red-winged Blackbirds in the Marsh

In spring, across most of North America, wherever there is a reedy pond or marsh, male red-winged blackbirds announce their presence. A male leans forward, elevates his wings slightly to show off the brilliant red-orange patch of feathers on his shoulders, and emits a penetrating squawk. Each male is displaying on his territory, a small patch of water and reeds to which one or more females will be attracted. From his perch, each male can survey his entire territory and he can quickly fly to chase away an intruder that appears anywhere along the boundary. A territory contains sites where a nest can be built and it contains enough area to provide the insects that the male, his mate, and especially, their broods, will require. Males that defend high quality territories with good nest sites and abundant food may attract several females and thus will father several broods of chicks. Peripheral males on poorer territories near the marsh edge may attract only one female, or perhaps none. In red-winged blackbirds, males compete for mating opportunities by laying advance claim to the resources that females need to rear a brood of chicks. This is a very common sexual theme.

Disposable Soldiers

All species of ants live in colonies in which one or a few reproductive females lay all the eggs and most females are sterile workers that perform the tasks (nest construction, cleaning, brood care, food gathering, nest defense) needed to maintain the colony. All ants vigorously defend the nest against intruders and, in addition, many species defend a large territory surrounding the nest. An ant colony is particularly well suited to territorial defense because thousands of workers can be in many places at once. A territorial limpet may suffer considerable damage to its algal lawn, caused by an intruder, before it discovers and deals with the challenge. In contrast, an ant colony can have many individuals on patrol at all daylight hours. These individuals run back to the nest to recruit help if they detect an intruder, and they die fighting to protect it. From the perspective of the reproductive unit, the colony, the loss of a few hundred workers imposes a slight energy cost, which translates into little or no fitness cost. Ants can run a "police state" type of territorial defense at relatively low cost and hence, they have some of the most absolutely defended territories known among animals.

Original Speed Dating

The African kob is a medium sized (80–100 kg) antelope that ranges across a narrow belt of sub-Saharan Africa. Males defend territories, each territory a tiny circle, about fifteen to thirty meters in diameter. Territories are packed together into clusters of about fifteen, so each male has several territorial neighbors. However because the territories are so tiny compared to the size of the animal, a male can easily survey his entire territory boundary and jump to defend anywhere that a rival crosses the line. Kob eat grass, so such a small circle cannot support even a single animal. A male retains his territory for as long as he can, until he must leave to eat and regain all the fat that he has lost while defending his little patch. Clusters of tiny territories such as these are called leks, and they exist for one purpose only: mating. Females visit leks to find a sire, and the packed collection of males permits a quick comparison of what is on offer. A female kob typically enters a lek, moves among the territories, and then chooses one male as the sire of her next calf. The criteria that females use to make these comparisons are not completely understood, but one preference seems to be for males that occupy territories at the center of the lek.

Leks are not unique to kob; they are the main way that males compete for matings in about two hundred species of beetles, flies, butterflies, wasps, fish, amphibians, birds, and mammals. A recent human convention that embodies the same idea is "speed dating."

Nursery Defense

The garibaldi is a stout, brilliant orange fish that lives in rocky areas along the Pacific coast of North America. Males choose an area that they will defend and aggressively chase away all other fish and invertebrates. A male works diligently to pluck a patch of rock until it is completely bare. The bare surface is then colonized by one species of red alga, and the male continues to work hard to maintain a clean, pristine surface of this alga alone. The preparation and cultivation of an algal mat may require two years of steady work, and the intensity of a male's territorial defense is testimony to the magnitude of his investment. Garibaldi males are so focused on territorial defense that they only reluctantly back away, even from human scuba divers. A female garibaldi visits male territories, selects an algal mat that she likes, and deposits her eggs there for the male to fertilize. The female then leaves and the male continues to defend a territory around his nursery until the eggs hatch and the fry disperse. In this example of territoriality, not only is there defense of a resource, but the defender has created the resource.

Who Benefits?

Social dominance and territoriality, two of the most common expressions of social behavior, illustrate the rule of individual selfishness in animal social behavior. Behavior, like other aspects of phenotype, such as the length of the small intestine, or the shape of a molar tooth, evolves because of individual differences in reproductive success. Behavioral traits, over evolutionary time, become tuned to maximize individual fitness. Chapter 3 showed that fitness is a measure which shows the magnitude of an individual's lifetime reproduction compared to the magnitude for the individuals with the highest lifetime reproduction. Individual fitness can increase by an increase in an individual's reproduction, or by a lowering of the reproduction of others.

In a dyadic social interaction (one involving two individuals), there is an actor, that performs a motor pattern or linked series of

Table 9.3 In a dyadic social interaction, there are four possibilities for the joint fitness consequences of the act on the actor and the recipient.

	RECIPIENT FITNESS	
	+	−
ACTOR FITNESS		
+	COOPERATION	SELFISHNESS
−	ALTRUISM	SPITE

motor patterns, and a recipient, the individual that receives the consequences of the action. In such a social interaction, there are four possibilities for the combined fitness effects of the action on the actor and on the recipient (see Table 9.3).

The first general prediction that we can make from this table is that actions in the second row, those that reduce the fitness of the actor, should not occur. Evolution weeds out behavior such as walking up to a predator and exposing the throat, and the same logic applies to social behavior. Fitness reducing acts are always removed by natural selection. From the definition of fitness, we can make a prediction about the first row: cooperation should be rare and selfishness should be common. A dispassionate survey of how animals in nature treat each other confirms these predictions. Actions that look spiteful essentially do not occur. Actions that look altruistic are rare, but there are several well-known examples. Actions that look cooperative are rare, and actions that look selfish are the norm.

Altruism

The exemplars of altruism are the eusocial insects, the bees, ants, and termites that live in huge colonies, in which only one or a few females lay eggs and most individuals toil as sterile workers. The workers seem to display the ultimate form of altruism; they give up all of their own reproduction in service to the reproduction by another female. In *The Origin of Species*, Charles Darwin admitted that the eusocial insects were a challenge to his theory of evolution by natural selection. How could natural selection favor individuals that gave up their own reproduction? A century later, a plausible explanation of the apparent paradox appeared. The author was W.D. Hamilton, who we met earlier in this chapter as the author of *Geometry for the Selfish Herd*. Hamilton's primary insight was that in social groups, the degree of relatedness between two individuals is important in determining whether natural selection will promote an act that appears to be altruistic. The degree of relatedness, r, that Hamilton referred to has a precise meaning in evolutionary genetics. Relatedness between two individuals is the probability that those two individuals, at any given site on a chromosome, will have gene copies that both came from the same ancestor. Between two individuals r is the probability that for any gene, they share the gene identical by descent. Between a parent and its offspring r is 1/2, between full siblings also 1/2, between half-siblings 1/4, between a grandparent and a grandchild 1/4, between first cousins 1/8.

Using this definition of r, Hamilton defined a term that he called *inclusive fitness*. The concept of inclusive fitness asks us to acknowledge that offspring are simply one class of relatives, to which the parent is related by 1/2. During an individual's lifetime, as well as direct production of offspring, other offspring, of a variety of degrees of relatedness, are produced. A female may produce four offspring in her lifetime, but during that time her sister produces four offspring, her aunts produce twelve and her own children produce sixteen. Table 9.4 shows how we calculate her inclusive fitness.

Note that in this example, the female's direct production of offspring counts for only about one-quarter of her inclusive fitness. Hamilton proved mathematically that natural selection operates on individuals to make them try to maximize inclusive

Table 9.4 Example of how to calculate inclusive fitness. N = number, r = coefficient of relatedness.

RELATIVE	N	r	INCLUSIVE FITNESS INCREMENT
Offspring	4	0.5	2
Nephews and nieces	4	0.25	1
Cousins	12	0.125	1.5
Grandchildren	16	0.25	4
Inclusive fitness			8.5

fitness, not simply direct individual reproduction. Thus, when living in a group that contains relatives, an individual faces an allocation problem. To maximize its inclusive fitness, how should it distribute benefit (apparently altruistic acts) among those relatives?

Hamilton solved the allocation problem. First, for any dyadic interaction, we identify a benefit (B), which is the additional fractional number of offspring that the recipient of the act will produce as a consequence of the act. Next, we identify a cost (C), which is the decrease in the fractional number of offspring that the actor will produce as a consequence of performing the act. Hamilton proved that natural selection promotes the performance of the act, as long as $Br > C$.

This inequality is known as *Hamilton's Rule*, and it specifies the conditions under which behavior that looks like altruism will evolve. Hamilton's prediction was that this kind of behavior evolves when the altruistic acts are directed toward relatives. An individual animal is related to itself by $r = 1$, so Hamilton's rule for individuals in their interactions with nature reduces to the familiar $B > C$: the benefit of an act must be greater than the cost. If the individual is interacting with a relative, for example offspring where $r = 1/2$, then the benefit must be more than two times the cost. For many types of parental care, which make the difference between an offspring surviving or not, this is certainly true. Finally, at very small values of r, in a group of non-relatives, the ratio of benefit to cost must be so huge that there is no realistic circumstance that would ever promote altruism.

Several clear predictions emerge from Hamilton's Rule. First, wherever we see behavior that looks like altruism, it should be in groups of animals that contain close relatives. Second, animals should practice nepotism: when they have a choice, they should distribute altruistic benefit to relatives, and they should give benefit to relatives in amounts proportional to r. A corollary of the second prediction is that animals, especially those that live in groups that are a mix of kin and non-kin, should evolve kin identification mechanisms.

Biologists started to test these predictions in the mid-1960s, about twenty years after the appearance of Hamilton's paper. Shortly afterwards, Edward Wilson of Harvard University published *Sociobiology*, the first of several books that brought him to celebrity status. In the book, Wilson reiterated the claims of the previous generation of ethologists, who had argued that behavior was a proper subject of biological study. However, Wilson placed emphasis on the biological study of social behavior, and championed the approach of deriving hypotheses from evolutionary principles (as Hamilton had done) and then testing those hypotheses with observations and experiments. In the subsequent three decades, many biologists followed this approach, and we have learned a lot about the social behavior of animals in nature. Before this revolution in behavioral biology, the premise that animals in nature acted as if they were constantly asking themselves the question, "what should I be doing at this moment to maximize my inclusive fitness?" was just that: a premise. Now it is solidly established fact.

Animals sometimes act in ways that seem detrimental to themselves and beneficial to non-relatives. For example, alarm calls are common. Imagine a group of Columbian ground squirrels out foraging. A coyote appears at the edge of the colony. The first squirrel to see the coyote gives a short, high-pitched bark and runs to the mouth of its burrow. Or, it may run to the mouth of its burrow first, and then vocalize. All the other squirrels within earshot run to their burrows. Although the behavior looks like altruism, the costs are so small that they are almost impossible to measure, and the benefits to recipients are similarly difficult to detect. Altruistic acts, that have real costs to donors and significant benefits to recipients, only occur among relatives. When we consider acts that have clear costs, such as delivering milk to non-offspring (milk is energetically rich and very expensive to produce), the pattern of kin-only benefit is obvious. In only twenty or so mammal species do mothers act to permit the delivery of their milk to young that are not their own. In all these species, the mothers live in stable social groups that comprise close relatives. In species that have social groups, such as herds of wildebeest or beach throngs of fur seals, mothers have evolved fail-proof discrimination mechanisms that allow them to permit only their own young to suckle. The altruistic provision of significant benefit at significant cost always turns out to be nepotism: simply another example of selfish individuals working to maximize inclusive fitness.

This discussion brings us back to the question of the eusocial insects. Do the sterile workers conform to Hamilton's Rule? Hamilton concluded that they do, and that this is so because of haplodiploidy. The *Hymenoptera* (bees, wasps, ants) have haplodiploid sex determination. Males are haploid (there is one copy of each chromosome per cell and thus all sperm are genetically identical). In addition, mating is decoupled from fertilization. A female mates, and stores the sperm in her spermatheca. When she lays an egg, she can decide whether to fertilize it. An unfertilized, haploid egg develops into a new haploid male. A fertilized, diploid egg develops into a female. This method of sex determination probably evolved in parasitoid ancestors of the *Hymenoptera* as an adaptation that allowed females to control the sex ratio of a clutch when local competition for mates was variable. Hamilton pointed out that haplodiploidy in eusocial *Hymenoptera* also creates unusually high r between workers, all of which are sisters. In the extreme case, if a queen honeybee has mated with a single male, any pair of workers will share 100% of their paternal genes identical by descent (because all the haploid sire's sperm were genetically identical), and on average share 50% of their mother's genes identical by descent. Honeybee workers are thus related to each other by r = 3/4. If a worker chose to mate and to lay her own eggs, she would be related to those daughters by r = 1/2. Even when a queen mates with several males (which is typical), the degree of relatedness among workers remains greater than 1/2. Ultimately, the reproductive output of a hive is a new queen, who will be a sister of the workers. Hamilton pointed out that because of haplodiploidy, a worker can achieve greater inclusive fitness by laboring to see that a sister becomes a new queen than by directly reproducing herself.

Recently, several researchers, including Edward Wilson, who study the social insects, have challenged Hamilton's hypothesis, noting that it is neither necessary nor sufficient to explain the distribution of eusociality. It is not necessary, because eusociality is common in another insect group, the termites, that do not have haplodiploid sex determination, and because eusociality occurs sporadically in other animals, including a mammal, the naked mole rat. It is not sufficient, because most species of the bees and wasps are not eusocial. These researchers are correct to point out that haplodiploidy does not create an automatic path to eusociality. However, given the fact that most of the eusocial species are *Hymenoptera*, it is likely that Hamilton identified one of the main characteristics that predispose an animal group to evolve toward eusociality.

Cooperation

Finally, we come to the most confusing and ambiguous cell of the fitness outcomes matrix shown in Table 9.3. It is the upper left cell, labeled "cooperation." Cooperation indicates social interactions in which both the actor and the recipient gain in fitness. Theoretically, interactions that represent this category are unlikely, because of the nature of fitness. Fitness is defined as relative reproductive success, meaning that an individual can increase its fitness by increasing its own reproduction, or by decreasing the reproduction of others. Thus, interactions that at first are cooperative will tend to slide, evolutionarily, to the right, toward selfishness.

Alpha and Beta Manakins

A good example of what most behavior that seems to be cooperation between non-relatives looks like occurs in lance-tailed manakins. The manakins are a family (*Pipridae*) of neotropical fruit-eating birds. Most species have a lek-based mating system. On their small territories, males try to attract mates by acrobatic displays that involve short flights, hops, and very unusual movements of the wings to produce buzzing or snapping sounds.

Lance-tailed manakins have mate attraction displays of this sort, but they add an element. Unrelated males often display as pairs, performing coordinated song and dance. For example, in the leapfrog dance (Figure 9.1c) two males stand in line on a cleared section of a branch, called a display perch. A female stands facing them. The male closest to the female leaps vertically, gives a call, and hovers briefly before landing just behind his takeoff point. While he hovers, the male behind him shuttles forward beneath to take up the forward position. He then leaps upward and the males continue leapfrogging and calling for ten to forty-five seconds. Although the two males seem to work equally hard in display, they do not share equally in rewards. The two males are

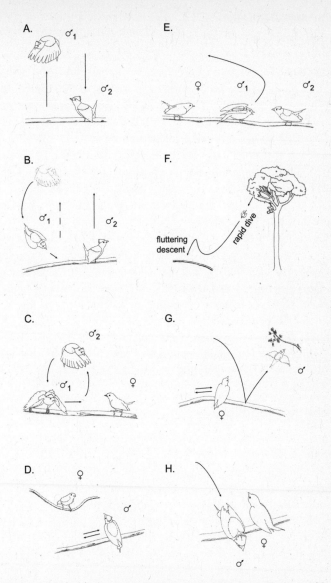

Figure 9.1 Solo and cooperative displays of male lance-tailed manakins. From: DuVal, E. H. (2007). 'Cooperative display and lekking behavior of the lance-tailed manakin (*Chiroxiphoa lanceolata*)'. *The Auk* 124, 1168–1185. Original artwork by Julian Kapoor.

designated as alpha and beta because one of them (the alpha) fathers all the chicks produced by copulations in the display area of the two males. Emily DuVal, who discovered that alpha males have exclusive mating rights, also tracked the fates of several beta males. She found that beta males sometimes, but not always, ascended into alpha status when the alpha disappeared. The coordinated courtship looks like cooperation. The beta male helps out his partner and in return gets a chance to become an alpha. Also, it may be that the practice in displaying helps him to become more proficient.

However, there are still several pieces of evidence missing. First, not all beta males ascend when their alpha disappears. Second, some alphas have no beta and they appear to have no difficulty in attracting females. Third, DuVal does not know at present whether pairs of males are better at attracting a female or at persuading her to mate. These are questions that she is currently investigating. At present, it is not clear whether the alpha male receives any benefit from having a beta male alongside. Considerations of this sort make the study of cooperation challenging and confusing.

Penguin Huddling

True cooperation, however, appears to occur in Emperor Penguins, large birds native to Antarctica that have a bizarre incubation routine. Pair formation, mating, incubation, and chick rearing occur on land. After a pair mates in the Antarctic fall, the female lays a single egg. To survive, the egg must be kept off the ice and remain hot. The male positions this egg on top of his feet and covers it with a fold of skin and feathers. The female then leaves for about 110 days, to feed in the nutrient rich waters off the Antarctic coast. The male incubates the egg, balancing it on his feet throughout the Antarctic winter. Because the male has no chance to feed during this time, and because he must endure very cold temperatures and howling winds, his ability to keep the egg alive throughout its development depends on a careful metering of the rate at which he burns his stored fat. If he runs out of fat before the end of incubation, he will not be able to maintain the proper temperature. The egg will die and he may as well.

Males meet this energetic challenge by cooperative huddling. For about a third of each day, and especially when wind speed increases, the males group together into very tightly packed huddles of hundreds to thousands of individuals. Because the males are crammed together (about twenty-one penguins per square meter), individuals in the interior of the huddle experience ambient temperatures that are much warmer than the outside air temperature. When the prevailing air temperature is −40°C, the temperature surrounding a bird near the center of the huddle is balmy, close to + 37°C.

Individuals join huddles from the rear and eventually exit from the front. A joining individual does not try to force his way into the interior. He simply plasters himself against the rear-most rank and waits until group movement and penguins joining behind him carry him into the interior. The movements within these huge, tightly packed groups are accomplished by coordinated stepping. Each male takes a tiny step − five to ten centimeters − forward at thirty to sixty second intervals. This stepping sweeps through the group in a rapid wave. Thus, the group can remain tightly packed (moreover, the movements act to promote the tightest possible packing, similar to the action of tapping on a jar of flour to increase density), but can still move.

By joining a group and participating in coordinated stepping, a male can lower his metabolic rate by 20% to 30% and so burn about half of the fat he would if he stood alone. Thus, he can survive the 110 day fast, and keep the egg balanced on his feet warm and alive. By cooperating in this way, the males create a fitness advantage that would not exist if they did not cooperate. Importantly, all males share equally in the benefit, and there is really no way to cheat.

Figure 9.2 Emperor penguin males huddling during the Antarctic winter. Photo courtesy of André Ancel, CNRS/IPEV.

Most apparent examples of cooperation turn out to be selfishness in disguise, or, like the coordinated dances of alpha and beta lance-tailed manakins, ambiguous. However, emperor penguins show that true cooperation can arise, when the conditions are just right.

10
Communication

Most animal species, whether or not they live in social groups, have behavior that exists solely to alter other individuals' behavior. Most communicative acts (behavior that animals perform with the goal of altering the behavior of others), are derived from other kinds of behavior, including respiratory movements, urination and defecation, thermoregulatory movements, protective movements, and intention movements.

In an intention movement, an animal begins to perform an action and then stops. For example, if you walk toward a robin that is feeding on a lawn, at some point the bird will stop feeding to watch you. If you walk closer, the bird will crouch and elevate its wings slightly. This is a flight intention movement. If you walk still closer, the intention movement will turn into an upward jump and flight. If you back away, the robin will rise from its crouch and re-tuck its wings. An intention movement is a way that an animal prepares to perform an act, but without committing the animal to the act. The amazing insight that the postures and actions used in communication are derived and modified from other types of behavior, such as protective reflexes or intention movements, came from the Nobel laureate, Niko Tinbergen.

Tinbergen's hypothesis concerned province number four – evolutionary history – of his four questions. Imagine that an individual is in a social situation in which its fitness could be enhanced by altering what another animal was likely to do. Perhaps it is in a standoff with a territorial neighbor. If the animal performs an intention movement to strike the other and this causes the other to back away, natural selection will favor individuals that are increasingly likely to perform the intention movement in that circumstance. Further, natural selection will favor the modification, perhaps exaggeration, of the movement, if this modification results in a more certain response from the recipient. Tinbergen referred to the evolutionary process in which an initial act becomes modified to have more reliable signal value as *ritualization*. Tinbergen further proposed that eventually, the new signal would no longer require the original brain activity that created the intention movement; it would acquire new control. Tinbergen referred to this change as *emancipation*.

It should not be too surprising that communication is about an individual animal selfishly trying to change what another animal does. In some instances, the sender of the communicative signal and the receiver of the signal have at least a partial common goal. In other instances, the sender and receiver are in direct conflict. As an illustration of these principles, as well as of methods that researchers use in the study of animal communication, consider a recent study of signaling by nestlings to parents in reed warblers, and the exploitation of this signaling system by a nest parasite, the common cuckoo.

The reed warbler is small, drab, insect-eating bird that nests, as the name implies, in reed beds in Europe and western Asia. The parents construct a basket nest that is attached to vertical reed stems above water, and lay about four eggs. As with most perching birds, after the eggs hatch, both parents work hard all day long to bring food to the waiting mouths of the nestlings.

The nestlings signal their hunger by throwing their beaks wide open to display the bright yellow skin inside the mouth (the "begging gape") and giving the cheeping begging call. To study this communication system, the researchers first tested the hypothesis that the begging signals of nestlings were an accurate reflection of their need for food. Researchers removed nestlings temporarily from nests, fed them to satiety in the laboratory, and then recorded the begging gape area and the call rate of the four nestlings at different intervals after feeding. They found that both the gape area and the call rate increased as the elapsed time since feeding increased. The nestlings signal increasing need for food by increasing their tendency to display the begging gape and by calling at faster rates. Having shown that this composite signal was a fairly accurate, honest indicator of offspring need, the researchers then turned to the responses of the parents, the recipients of these signals.

In nature, the researchers studied nests that contained two, three, or four nestlings, and attached a small loudspeaker to the side of each nest. When parents visited the nest, they found either their normal brood, the brood plus the broadcast vocalizations of one additional chick, or the brood plus the broadcast vocalizations of four additional chicks. The researchers then watched the nests to measure the rate at which parents brought food. They found that the feeding effort of parents (the amount of food that they brought to the nest) was largely explained by begging gape area (more chicks in the nest, more of them with mouths open) and by the begging call rate (the total number of calls per second). Just as the chicks produce a composite signal of need, so parents appear to integrate both signals to adjust their feeding effort.

This honest signaling system is exploited by the common cuckoo, an obligate brood parasite that commonly parasitizes reed warbler nests. Obligate means that the only way for a cuckoo to reproduce is to lay an egg in the nest of another species, tricking the host parents into caring for the chick. A cuckoo female surreptitiously watches a reed warbler nest, waiting until eggs are present but the parents are temporarily absent. The cuckoo female then flies to the nest, removes an egg, lays her own egg, and flies away. The entire duration of this stealth visit is about ten seconds. The cuckoo egg tends to hatch first, and the cuckoo chick, immediately after hatching, shoves all other eggs and nestlings out of the nest.

The growth rate of the cuckoo chick, in grams per day, is a little less than that of four reed warbler chicks, so the reed warbler parents are capable of delivering enough food to meet the demands of the parasite. To keep the food coming, the cuckoo chick needs to deliver signals that are equivalent to those produced by four reed warbler chicks. At first, the gape area of the cuckoo chick is only a little less than that of four reed warbler chicks, and the cuckoo chick produces begging calls at about the rate that would be produced by four reed warbler chicks. After a few days, the gape area of the single cuckoo chick becomes progressively smaller than that of four reed warbler chicks. The cuckoo chick compensates for this by elevating the begging call rate to a greedy, insistent rate of ten to twenty calls per second. This constant yammering by the parasite chick is sufficient, along with its gape area, to induce parental feeding at the level appropriate for four reed warbler chicks.

Aggressive and Defensive Signaling

Many signals between members of a species occur not in a mostly cooperative circumstance, as in the communication between parents and offspring, or between potential mating partners, but in a competitive context, in which the interacting individuals are in conflict. Conflict occurs because essential resources, those needed for survival and reproduction, are usually limited. The essential resources are food, nesting or refuge sites, and mates. As discussed in the previous chapter, animals often attempt to lay advance claim to resources by social dominance or defense of a territory. Thus, some of the most common types of signals are those involved in the assertion of dominant status, the signaling of subordinate status, and the advertisement of territory ownership.

Signals that assert dominance tend to be derived from attack intention movements. The dominant individual assumes an upright posture, facing directly toward the subordinate, and makes a gesture that is an intention movement to strike or is a subtle derived version of such a movement. Thus, in monkeys and apes (including humans), we can read dominance status by posture alone. Signals of subordinate status are derived from protective movements: the subordinate animal crouches, or hunches, or lowers the head, and turns away.

This kind of signaling, where animals contest dominance status or, more directly, ownership of a resource, tends to have elements that one could call dishonest. The false signaling of body size is very common. In aggressive displays, individuals adopt postures that make them appear larger, and they often have physical structures to enhance the illusion. For example, male bison increase their apparent size during the mating season with a huge crown of hair on top of the head, large dangling beard, and pantaloons, big tufts of hair on the lower legs. One of the male threat postures is to face directly toward his opponent, with the head slightly lowered. In front view, these three hair accessories greatly increase the male's apparent size. After the mating season, the males lose most of the extra hair. More rarely, animals succeed in transmitting false information about their abilities, as well as their size. However, we only know of a few examples, because in contests, assessment of rivals is usually thorough.

Signals that advertise territory ownership are common because the alternative is continuous territory patrolling and boundary defense. If an animal can simply give a signal that an area is occupied, it may be able to devote more time and energy to other important tasks. Vocal signals of territory ownership are common, because a vocal signal is broadcast. Although birdsong is well known, vocal advertisement of territory also occurs in mammals. Wolf howling and lion roaring are familiar examples, but there are others, less famous but equally thrilling. The tropical forests of Southeast Asia were the evolutionary home to a group of apes, the gibbons. These primates are forest tree specialists. They use their long powerful arms to swing from branch to branch or to sail across gaps of several meters. Gibbon social groups, a mated pair and their recent offspring, defend a territory that contains sufficient resources, especially fruit trees, to provide adequate food. Because gibbons can move rapidly through the forest canopy using their spectacular swings and flying arcs, they can physically patrol a territory and defend its boundaries. But gibbons also announce territory ownership with loud, beautiful songs. The principal singers are the females. Their song is a series of pure, rising tones that are spaced at about half-second intervals at the start and become packed together somewhat like a bird's trill toward the end. In addition, most species sing duets, in which the male sings either a coda to the female's song or a more complex counterpoint. The gibbons broadcast these loud haunting songs from the treetops, announcing their presence well beyond the boundaries of their territories.

Scent Marking

Although some mammals advertise occupancy of a territory by vocalization, the more common method is scent marking. Nearly all mammals possess scent glands, patches of skin specialized to produce odor compounds. Scent glands are either modified sweat glands or modified sebaceous glands – the oil producing glands at the base of hair shafts. Mammals also display specific scent marking motor patterns, which are designed to deposit material from the scent gland on an object in the environment. Chromatography shows that scent gland material contains dozens to hundreds of specific chemical compounds. Thus, individual or group identity can be coded by variation in the proportions of these compounds. Also, mammals commonly scent mark using urine and feces, other sources of socially significant odor. By scent marking within a territory or home range, a mammal can create a zone where the odor is familiar and thus even on a dark night (most mammals are nocturnal) have a sense of whether it is home or not. In addition, although scent marks never act like an absolute "odor fence," they do tend to slow an intruder down and to make it more nervous.

Scent marking is also used in some species to coordinate reproduction. This effect is well known, for example, in the house mouse, a superb mammal colonist. Dispersal and colonizing ability is supported by an odor signaling system that coordinates reproduction. A male house mouse has a long, thin prepuce, which he drags across the ground, depositing a thin streak of urine. Socially dominant males scent mark like this several thousand times a night, saturating their territories with their own scent. Female mice need to smell an odor in male urine to start reproductive cycling. In the laboratory, exposure to tiny, microliter, amounts of male urine will start mature females cycling and will accelerate the onset of reproductive maturity in young females. Male urine marking, in turn, tends to increase in the presence of odors from female urine. This reciprocal signaling system helps individuals to be effective colonists. A female that has left her natal area to disperse to a new habitat does not waste energy in unproductive estrus cycles. She waits until there is a reliable indicator that males are present. A dispersing male, detecting the odor compounds indicative of female urine, elevates his rate of scent marking to provide a signal that will produce reproductively active females.

The production, detection, and analysis of scents are central to the lives of most mammals. Besides scent marking to anoint personal space and to coordinate reproductive timing, mammals use odors to identify individuals, to detect levels of genetic similarity to others, to evaluate potential mates, to maintain group cohesion, to signal the presence of a predator, to attract mates, and to evaluate the reproductive readiness of potential mates. Most mammals have levels of chemical sensitivity that are far greater than humans. Part of their reality is an odor world that we are only dimly aware of. To study odor communication in animals, we humans require an organic chemistry laboratory, a gas chromatograph, and other complicated equipment, just to describe the nature of the signals and to begin to alter them in the way that early ethologists studied visual signals with paper cutouts.

The other animal group in which chemical communication is nearly ubiquitous is the insects. In contrast to mammals, with scent glands that produce a complex mixture of smelly compounds that produce general responses in recipients, insects often produce pheromones, single chemical compounds, or simple mixtures of compounds, that elicit specific behavioral responses from recipients. For example, in many species of ants, individual foragers lay down odor trails that lead from the nest to a food source that they have found. New foragers, exiting the nest, find the odor trail and follow it to the food source. If the recruited worker finds food there, it also lays down a scent trail as it carries the food back to the nest. In this way, a number of workers is recruited sufficient to quickly dismantle the food source (perhaps a dead grasshopper) and transport it back to the nest. The trail pheromones are quite volatile; they evaporate after a few minutes. Thus, old, useless odor trails do not persist. In the fire ant, a South American species that has become an invasive pest in North America, the principal trail pheromone has been identified. If one coats the tip of a glass rod with a drop of this compound (a drop is a fantastically huge amount of this substance to the ants) and holds the rod over the colony entrance, one can draw out the entire colony, as individuals climb on top of each other to follow the odor.

Besides trail marking, ants use chemical communication to coordinate most activities within the colony. In their magnificent summary of the biology of ants, Bert Hölldobler and Edward Wilson state that "The typical ant worker is a walking battery of exocrine glands ..." An exocrine gland produces compounds that are released into the environment. The responses to specific gland products are so rigid and predictable that several slave-making species of ants have evolved chemical tactics. They produce the pheromones of another species and release them when on slave raids, rendering the hosts' defense ineffective in one way or another. In at least two cases, the aggressive pheromone release causes members of the raided colony to attack each other while the slave makers go about collecting eggs or larvae that will become new slaves.

Chemical communication is also common in the non-social insects. In many moth species, females that are ready to mate release an attractant pheromone. Males detect the pheromone using their sensitive antennae. Sensitivity is so extreme that in some species, one molecule of the pheromone is sufficient to induce a neural response from the antennal receptors. A male flies toward the source of the odor until he locates the female, and courtship begins. Often, male courtship involves further chemical communication, in which the male turns out the brush-like scent organs near the tip of his abdomen and sweeps them back and forth across the female's head.

The sex attractant response of moth males also renders them vulnerable to human pest control measures that exploit the communication system. Agricultural research workers identify the chemical compound that is the pheromone of a pest species, synthesize large amounts of the compound, and place generous amounts inside box traps. The traps quickly become stuffed with males.

Responses to Predators

In prey species, individuals often produce communicative signals, using postures, movements, vocalizations, and scents when they detect a predator. After giving the signal, individuals may approach the predator, while continuing to signal. Typically, several other prey individuals join in the approach. The initial signaling is sometimes involved in recruiting participants. This group approach toward a predator is called *mobbing*. In birds, agile prey species may dive at the predator and occasionally strike it.

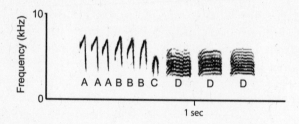

Figure 10.1 Sonogram of the chick-a-dee vocalization. The notes labeled A, B, and C are the "chick" part of the call, and the notes labeled D show the nasal "dee" sounds. Image courtesy of Mike Baker.

A common North American bird, the black-capped chickadee, lives in stable social flocks after the breeding season throughout winter. Birds recruit others to join in mobbing using a "chick-a-dee" call when they detect a perched predator close to a flock. In one study, researchers presented chickadees in an outdoor aviary with living predators in an adjacent cage. The predators varied from large, slow, birds, such as great grey owls, to small agile ones, such as kestrels. The more dangerous predators elicited more calls, and more D notes per call.

Other researchers did not find more D notes in the more dangerous situation. In one study, researchers presented a taxidermic mount of a prairie falcon to single chickadees at two distances: close (one meter away) or far (six meters away), and recorded the vocal response. In this study, the number of D notes per call did not differ between close and far trials. Currently, there is disagreement on how the proportions of A, B, C and D notes are used to convey meaning. One way to resolve the differences of interpretation would be to record calls from all individual chickadees in a flock in nature that was interacting with a predator and to simultaneously record the reactions of all other birds and of the predator. At present, the technology to do this does not exist.

Mobbing is an example of communication between, as well as within, species. The goal of the behavior is to cause the predator to go away, so that the prey individual can resume its normal activities. A particularly clear illustration of the fact that mobbing signals have evolved to influence the behavior of predators occurs in California ground squirrels. These burrowing rodents had a long coevolutionary history with rattlesnakes, and they evolved a number of defenses. Adult squirrels are resistant to rattlesnake venom and they can jump rapidly to avoid snake strikes. However, their pups are slower and do not yet have the chemical resistance. They need protection when a rattlesnake appears. An adult parent approaches a hunting snake and stands with its tail held upright, with the hair erected. The squirrel waves its tail from side to side (called tail flagging) and may kick dirt in the snake's face. This behavior slows the snake's approach or causes it to go away. A channel of communication that is specifically directed at the snake in these encounters is infrared radiation. Rattlesnakes have infrared radiation detectors on their faces (in effect, they have a pair of poor resolution auxiliary eyes, tuned specifically to electromagnetic radiation in the infrared range). When a California ground squirrel tail flags at a rattlesnake, a significant amount of infrared radiation glows from its tail. When the squirrel tail flags at a gopher snake (which does not have the infrared detectors) the tail remains cool. California ground squirrels use a signal that is designed specifically to communicate with rattlesnakes.

Courtship

A final major category of communicative acts involves courtship; the behavioral signals that precede and coordinate copulation. Broadly, courtship comprises two kinds of signals. First, there are the signals that are involved in attracting a mate, as in firefly flashing, cricket chirping, and frog choruses. Although most birdsong acts as advertisement of territory ownership, in many species males occupy territories first and song initially has two functions. Song helps to keep other males away, and it is broadcast to attract a female mate.

However, despite these well-known examples of acoustic advertisement for mates, the more common mate attraction signals are chemical. Chemical sex attractants have been identified in more than 1,600 species of insects. In many mammal species, females, as they approach estrus (coordinated ovulation and behavioral receptivity to copulation) release sex attractant compounds in urine, or as volatile scents from the vagina.

The effects of the signals are to bring males and females of the same species together. However, mate attraction alone does not guarantee copulation. Usually, mate attraction is followed by a further round of signaling, which may include movements, sounds, and scents. The purpose of this second round of signaling is twofold. First, it is a check on species identity. One of the worst mistakes that an animal can make is to waste a reproductive opportunity by mating with an individual of another species. Second, the added round of signaling is used to evaluate the suitability or quality of a potential mate.

In long-lived animal species in which individuals mate for life (many seabirds, for example), the outcome of this second signaling round is critically important for an individual's lifetime fitness. Typically, animals with these life history characteristics have a relatively low annual reproductive output, and the one or two young from each breeding season do not always survive. To accumulate fitness, an individual must make a solid effort at reproduction every year, over many years. In seabirds, as in most birds, the hatched young have huge energy demands that can only be met by the combined efforts of both parents. When a young adult bird is about to commit to a pair bond with another individual, that will be its reproductive partner for perhaps thirty years, it is important to make a good choice.

An example of such a species is the Laysan albatross. Individuals that are shopping for a mate perform an elaborate "dance" with the potential life partner. In a dance, both individuals face each other and display a rich set of motor patterns that one researcher named: 1) bill under wing, 2) scapular action, 3) sky moo, 4) sky call, 5) air snap, 6) sky snap, 7) head flick, 8) head shake and whine, 9) stare, 10) stare and whinny, 11) rapid bill clapper, 12) eh-eh bow, 13) bob, 14) bob-strut, 15) bow-clapper, 16) bill touch. The dance is coordinated in some way, so that while one partner is performing one set of actions, the other predictably performs another set. The birds bob up and down, clap their beaks loudly, move the beak under the wing or to the back and clap the beak, shake the head, throw the beak to point vertically upward and give a call; in other words, perform a rich set of ritualized actions that are seen in no other context. Exactly what motions a potential partner attends to is not known, but it is common for one partner to break off and walk away. It has seen something that indicates unacceptable quality in the dance partner, and it will search for another possible mate.

11

Sexual Behavior

The northern elephant seal gets its name from the males' fleshy, pendulous snout, which looks like a short version of an elephant's trunk. The snout, somewhat like the vocal sac of frogs, acts as a resonating chamber, to alter the amplitude and frequency characteristics of male vocalizations. Males produce their loud, breathy, flatulent vocalizations from their beachfront territories, advertising ownership, in the seal equivalent of birdsong.

Northern elephant seals spend most of their lives at sea. In the rich waters off the northwest coast of North America and up into the Gulf of Alaska, they find abundant squid, the main item in their diet. Seals are mammals, mid-way through the process of evolving from terrestrial life to aquatic life. They cannot give birth or care for their young in the water; seals must haul out onto land or ice to give birth to their pups and suckle them.

Most seals, including elephant seals, have a postpartum estrus. They copulate and initiate a new pregnancy almost immediately after giving birth. Favorable beaches on islands where females can come ashore to give birth away from terrestrial predators are rare. Thus, many females tend to appear at one spot. The concentration of receptive females in space and time sets the stage for competition among males.

In elephant seals, as in nearly all mammals, all the energy expense of producing young is borne by the females. A female grows the fertilized egg in the uterus, feeding the developing fetus directly from her own blood, and then continues to feed the young after birth in the energetically very expensive process of lactation. Males simply fertilize the egg. Thus, a male's lifetime reproductive success is determined solely by the number of eggs that he fertilizes, that is, the number of sexually receptive females with which he copulates. A male can elevate the number of copulations that he achieves in two ways: he can attempt to sequester females and he can attempt to keep other males away from them.

Elephant seal males strive for these goals by arriving early on the island beaches where the females will later come ashore. Mature males set up territories right at the water's edge, so they can intercept and corral females as they come out of the surf. A male with one of these beachfront territories assembles a harem of several dozen females. A female finds these males useful, because they defend a territory where she can give birth in relative peace, not bothered by throngs of suitors.

Males defend their territorial boundaries from neighboring males and from younger males that attempt to enter. A male's defense of his territory involves guttural song, threat postures, and fighting. Fighting males rear up facing each other and slash downward at the opponent's face and neck with their powerful canine teeth. Although males have specially thickened and calloused skin in these areas, they still become cut. Two adversaries in a prolonged fight will have blood streaming down their necks. The fighting blows are also a kind of clubbing, in which the force of each male's mass is brought downward on the opponent. Because body mass in part determines the outcome of these fights, mature males are large; their body length is about five meters and body mass is about 3,000 kilograms. Female body length is about three meters and body mass is about 900 kilograms.

A mature male is about three times the size of a mature female for one reason. Male size determines the outcome of fights, and fights determine whether males retain territories and get to copulate. The utter dependence of male reproductive success on male mating success is the reason for the male proboscis and the vocalizations that it helps to make, for male territoriality and defense of females, for male pugnacity and willingness to fight, and for complete disregard for the welfare of pups or the mating preferences of females. When a male, resting near the center of his territory, detects another male encroaching the boundary, he runs, as best as a three tonne animal that is designed for swimming can, toward the intruder. A running seal flexes its spine up and down to throw its body into rippling waves and flows across the ground, using its flippers for support. As the territorial male charges toward the intruder, any females in his way scatter or are thrown to the side. The male simply runs over small pups that are unlucky enough to be in his path. Under his crushing weight, they are injured; often killed.

When a male detects, by odor, a postpartum estrus in one of the females in his harem, he lies across the female, pinning her to the ground with his weight while he probes to insert his penis. The female cannot get away. Her only option in this situation, if she finds the male to be an unacceptable mate, is to give a protest vocalization, which may draw in other males, forcing the male on top of her to get off and fight.

Sexual Selection

The evolutionary process by which sex differences in body size and behavior come about in elephant seals is called *sexual selection*; a term coined by Charles Darwin in his first book. There, he presented and evaluated the idea that the principal cause of evolutionary change in species was natural selection. Natural selection is the process by which species change genetically to become increasingly better fitted to their environments. In *The Origin of Species*, Darwin thoroughly considered all the possible challenges to his theory. One obvious challenge is that in many animal species, males and females differ in aspects other than their genitalia. Within a species, males often are larger than females, they possess odd structures, such as the elephant seal proboscis, or the horns and antlers of most hoofed mammals, and they are easily motivated to fight. How can it be, Darwin realized, if the environment defines a single best optimum for a species, a form and function that makes individuals as good as possible at survival, growth and reproduction, that different forms, male and female, come about? If a body mass of 900 kilograms is best for a female elephant seal to gather resources and to convert these into offspring, why should males be three times larger?

Darwin concluded that there must be a process other than natural selection pushing males to evolve larger body size, weapons, and pugnacious dispositions. Darwin called this process sexual selection. He proposed that, because mating is a crucial required step in sexual reproduction, characteristics with the sole purpose of making individuals more successful at mating could evolve to extremes that were opposed by natural selection. At equilibrium, the fitness advantages of an extreme trait that creates greater mating success is balanced by the survival costs of the trait. Thus, in most species, we view the female condition as the optimum that is defined by the environment, and the male condition as somewhere off that optimum. The distance that males are away from the environmental optimum is set by the importance of male mating success for male reproductive success.

Polygyny

Another characteristic of species, such as elephant seals, in which males are specialized to compete for mating, is the substantial variance in male fitness. In a typical year, most breeding age males do not reproduce, and within the small fraction of males that do, there is a pronounced skew: a few males sire many offspring, and many others sire only one or two. In contrast, just about every female mates, and in the subsequent year gives birth to a pup. Thus, although the average reproductive success of males and females is equal (it has to be: one male and one female are needed to make each pup), the variance in male reproductive success is large and the variance in female reproductive success is small. This situation, where variance in male reproductive success is greater than variance in female reproductive success, is called *polygyny*.

In polygyny, males tend to be larger than females, to have weapons of some sort and the behavior to aggressively use those weapons. They employ various ways, including territoriality, to establish harems, groups of females that they attempt to hold together and defend against other males. They are behaviorally focused on copulation and do not form pair bonds with females. They are either indifferent to juveniles, as is true for elephant seal males, or they practice infanticide, as is found in species such as langurs or lions, where there is a turnover of the single males that defend female groups. When females mate more than once per receptive period, males have adaptations to deal with sperm competition.

Sperm competition refers to the fact that reproductive competition among males may not end at copulation. Until an egg is fertilized, there is an opportunity for one male to displace another. Male adaptations to deal with sperm competition are quite varied. In some species, males practice mate guarding after they copulate. They hover close to the female and do not permit her to go out of their sight until enough time has passed to ensure fertilization. In other species, males insist on frequent copulation. In the giant water bug, *Abedus herberti*, a male copulates with a female and then permits her to glue eggs to his back. While the eggs develop, the male does not feed, and moves in specialized rocking motions to ensure proper aeration of the eggs. Females may mate with several males, and thus a male is at risk of accepting and caring for eggs that he did not fertilize. Experiments in the laboratory show that when a female has mated several times, the male that mated with her most recently is the sire of the next egg that she lays. Thus, male giant water bugs allow a female to glue no more than three eggs to their backs before requiring another copulation. In one recorded instance, a pair of bugs took thirty-six hours to transfer 144 eggs to a male's back, and during this interval, copulated over a hundred times. In other insect species, male genitalia come equipped with structures that act like scoops, to remove another male's sperm. Similar flushing techniques are known in some birds and mammals. Male seminal fluid may contain compounds that, in the vagina, set up like glue, forming a rubbery plug that blocks additional sperm. In other species, seminal fluid contains compounds that are toxic to subsequent sperm. Finally, males often compete in the number of sperm that they ejaculate. Suppose that two males mate with a female, and one male ejaculates 200 million sperm, and the other male 800 million. If there are no other sperm competition mechanisms operating, the second male will fertilize 80% of the eggs. For this reason, males in polygynous species often have very large testes, ejaculate prodigious numbers of sperm, and can do so repeatedly, with only short periods in-between.

Another evolutionary consequence of polygyny is the existence of alternative reproductive tactics in males. The most common alternative reproductive tactic is one that is generally described as "sneaker." Most species of salmon have an alternative male form, a jack, which is a sneaker. Most male and female salmon spend two to several years at sea, feeding and accumulating the resources needed for a single massive bout of reproduction. At the spawning site, females dig shallow nests in the gravel to receive their eggs. Males defend these sites, and the females, with threat postures and fighting. Males are larger than females, and their jaws have developed an enlarged, hooked shape that resembles tin snips. (The genus name of the Pacific salmon, *Oncorhynchus*, means "hooked nose.") A male aggressively defends a female until she lays her eggs. He hovers close to her as she adopts the egg-laying posture, and when she releases her eggs into the nest, the male immediately covers them with a cloud of sperm. As in elephant seals, the success of a hooknose male is related to his size, strength, vigor, and fighting ability.

While they were at sea, the salmon females amassed the resources needed to produce as many high-quality eggs as possible. Males amassed the resources that would allow them, during the upstream migration, to transform into big, lethal fighting machines with huge testes. The alternative male type, the jack, does not stay at sea, but returns to the natal area after just one year. A jack is small, and it does not have the red color or the tin snip snout of hooknose males. A jack does not participate in the fights among males for ownership of a female and her nests. Instead, he lurks, usually in vegetation, close to a female and the hooknose male that is defending her. At the instant that the female releases her eggs, the jack darts in quickly and deposits his sperm. Then he swims rapidly away, trying to avoid the slashing jaws of the big hooknose male.

How does a young male make the "decision" to become a jack or a hooknose? The answer seems to be growth rate. If a male is healthy and growing quickly as a smolt, he is more likely to become a jack. At any given smolt age, a male that is smaller tends to choose to stay longer at sea and to return as a hooknose. These findings contradict the commonly held notion that the "sneaker" strategy represents the best option for a male in a compromised situation. In salmon, jacks are probably the higher quality individuals, and some recent evidence suggests that the behavior of spawning females is designed to favor fertilization of their eggs by jacks.

Monogamy

Most species of mammals are polygynous, due to the extreme asymmetry in parental expenditure in offspring that exists in mammals. Females provide all the energy and materials that are needed to produce an independent offspring, and thus in effect are a resource that males compete for. Given the absence of any kind of significant male expenditure in offspring, males become specialized at competing for mating.

In other animal groups, in which males put some energy into rearing offspring, we see what looks like monogamy. Most bird species appear to be monogamous, because the efforts of both parents are required to raise a brood. Two kinds of parental effort are needed in birds. First, the eggs must be incubated. They must be kept warm; nearly always, this is by a parent sitting on them. Constant incubation requires either that the parents take turns, or that one parent feeds the other. Then, after the eggs hatch, the energy demands of the chicks are gargantuan. The gaping chick beaks have to be stuffed with high quality food at very frequent intervals. Baby birds have huge energy demands because, first, they are endothermic (warm-blooded). Being small and lightly feathered, they must burn energy at a great rate to produce enough heat to maintain body temperature. Second, birds have very fast growth rates, which create another insatiable demand for energy. In most bird species, the efforts of two parents, working from sunup to sundown, are required to deliver enough food to bring the chicks to fledging; the point at which juveniles can fly from the nest and start to feed themselves.

Thus, in most bird species, individual males and females pair off. There is a courtship phase, which may be lengthy, during which the two birds in effect agree to be reproductive partners (recall the dancing albatross from the previous chapter). Both birds may help to build the nest. Quite often, both defend the territory boundaries. They copulate. The female lays eggs in the nest. Cooperative incubation and continued territory defense follow. The eggs hatch, and the parents either guard the nest and the nestlings, or forage, to ferry an unending supply of food into the mouths of the hungry chicks.

The reproductive biology of birds seems to require monogamy – equal variance in the reproductive success of males and females – because all males and females are members of a mated pair.

This was the consensus among ornithologists until the 1990s, when DNA technology progressed to the point at which genetic assignment of paternity became relatively easy. In species after species, the shocking news came out that the chicks in the nest often were not fathered by the social mate, the male working hard to do all that feeding: rather, a male in a nearby territory had fathered them. Ornithologists were forced to conclude that in the monogamous birds, many extra-pair copulations (EPC) happened. This was a shock, because field workers did not see these copulations. However, in some species, EPC rates are so high that none of the chicks a male cares for in his nest are his own. EPC is the rule, rather than the exception, and biologists now make a distinction between social monogamy and genetic monogamy. Most birds are socially, but not genetically monogamous, and the EPC occur surreptitiously.

It is relatively easy to see why a male might engage in EPC. If he can take a moment out of his busy foraging day for a quick copulation with the female next door, he can cheaply increase his fitness without having to care for the additional chicks he has fathered. Also, EPC are available before egg laying, and because all birds in a neighborhood are synchronized, EPC are available before a male has his own chicks to feed. However, the more time that a male spends out looking for EPC, the less time he can devote to monitoring his mate, and thus the greater his risk that she will have EPC with a neighboring male. At least for males, we can estimate a balance between EPC and mate guarding that maximizes fitness. However, in most species, what advantage females get from EPC is not that obvious. Researchers have proposed many hypotheses, but at present, there is not a conclusive answer.

Sexual Conflict

Nick Davies of the University of Cambridge studied the reproductive behavior of dunnocks (*Prunella modularis*), a small, secretive, brown bird, in the Cambridge University Botanic Garden. Davies was pursuing the question of why this species appears to show so much variability in its mating system. In the population that Davies studied, males and females independently defended territories, with varying amounts of overlap. On some territories, a single male and female practiced typical birdlike social monogamy. On other territories, a single male's territory overlapped with those of two females, and the male appeared to be mated with two females. On other territories, a single female was associated with two males. Finally, some territories had messy arrangements with two or more females and two or more males.

By monitoring the rates at which adults fed nestlings, and recording the fates (survive or die) of nestlings, Davies found that the sexes differed in what was optimal for reproductive success. For individual males, the greatest number of offspring fledged per season came about under polygyny, where a single male mated with two females and helped both to rear the chicks. For a female, this arrangement was not optimal, because she did not have the male's undivided feeding efforts and so the survival of her chicks was diminished. A female had the greatest reproductive success when she was associated with two males, that both fed the chicks. However, a male will feed a female's chicks only if he has copulated with that female several times, and thus has a reasonable chance of paternity. So a female with two males on her territory needs to copulate with both to be sure that both will feed the chicks. However, males also practice mate guarding, so the females are required to be quick and sneaky about soliciting copulations from male number two. If male number one sees his female copulating with male number two, he becomes less likely to feed the chicks.

A female in association with two males does best if she can elicit the services of both to feed the chicks, but the males try to evict each other, tend to suspend chick feeding if they do not get to copulate or if they see the female copulate with the other male, and insist on cloaca pecking, which causes the female to eject a droplet of the other male's sperm. When a male has two females on his territory, the females are hostile to each other, may interfere with each other's nests, and neither female gets the male's complete chick feeding attention. Each sex would prefer to have the exclusive reproductive services of two members of the opposite sex, and individuals of each sex attempt to prevent this from happening to them. This situation, in which the optimum conditions for the sexes are in opposition, is called *sexual conflict*.

Many reports of sexual conflict appear in the scientific literature. The grisly phenomenon of infanticide provides a particularly clear example. Infant killing by adult males is a reproductive adaptation. However, the female whose infant is killed does not benefit. The death of her infant is a fitness loss for her. The sexes are in fundamental conflict, and the evolutionary consequences of the conflict ripple outward. Infanticide following male takeover occurs in lions. The average size of groups of female lions, by some estimates, is larger than what is currently best to maximize the per capita intake rate of prey. Some researchers have proposed that these larger than optimal group sizes have evolved as a response to infanticide, because several females can cooperatively threaten a male away from a cub that he is trying to kill.

To humans, infanticide is a heart-rending adaptation, one showing that selection is indifferent to everything except fitness, allowing behavior that to us seems horrible to evolve. In other instances, the behavior that evolves from sexual conflict is more similar to Shakespearian comedy. One example occurs in a species of burying beetle. The burying beetles is a group of about seventy species that are specialists at using vertebrate (mostly mammal) carcasses as reproductive resources. In nature, big carcasses (such as a dead mouse or rat) are very valuable sources of energy, and many life forms, including bacteria, fungi, insects, and other vertebrates, are adapted to exploit them. Burying beetles have more elaborate parental behavior than most insects. To save a carcass for reproductive purposes, a pair of beetles dig beneath it, cover it with soil, clip off all the hair, treat the shaved surface with oral and anal secretions that soften the carcass and protect it from microbial attack, and mold it into a rough sphere. The beetles have thus prepared a rich ball of energy hidden in an underground brood chamber. The female lays her eggs and the larvae, when they hatch, either feed from an opening in the carcass that the parents have prepared, or the parents regurgitate partially liquified carcass material to them.

When a single beetle finds a carcass, he or she releases a pheromone that attracts a beetle of the opposite sex. Two partners are needed to quickly bury the find, as well as to produce fertilized eggs. If more than one male is drawn to the carcass, they fight violently until a single winner drives the other males away. Likewise, females attack and attempt to evict other females that show up. If the carcass is large enough, more than one female can rear a brood on it, but only at diminished output per individual. A female can rear more young if she is not forced to share. Here is where sexual conflict becomes apparent. A male will have higher reproductive success if two or three females come to the carcass. He can sire all the broods and collectively produce more offspring than if he remained monogamous. On a small carcass, which can support only one brood, a male remains monogamous. On a large carcass, the male tries to release the attractant pheromone, to draw in additional females. The female in the pair has a fitness interest in not allowing the male to advertise for more females.

This is the arena for sexual conflict in the burying beetle, *Nicrophorus defodiens*, which Anne-Katrin Eggert and Scott Sakaluk studied. In this species, males attempt to attract additional females by mounting an elevated spot, then performing a kind of headstand, to expose the terminal abdominal segment, from which pheromone is released. Females try to stop or prevent their mates from doing this by biting them, knocking them off balance, or jumping onto the male's back.

Figure 11.1 The male burying beetle on the right is trying to adopt the head stand posture to release attractant pheromone that will draw in new females. His mate, on the left, has jumped on his back to prevent this. The male is trying to kick her off. Photograph by Anne-Katrin Eggert.

Eggert and Sakaluk decided to measure the effectiveness of these female countermeasures in the laboratory, where pairs were presented with a rat carcass. To prevent the female from resisting advertisement by her mate, the researchers tied a tether of dental floss around the female's waist, so that she could move around the carcass, but not get off it. In the cage with the carcass, the researchers also provided a small rock that the tethered female could not reach. Then they measured how often the male managed to get into his headstand position and how long he was able to remain in that position, first when his mate was tethered, and then when she was not. With his mate unable to follow him to the little rock, the male spent significantly more time with his abdomen in the air, wafting his come-hither pheromone.

Female Choice

In his original presentation of the idea of sexual selection, Darwin intended to explain two kinds of traits that we see in nature. The first kind, as described, is that males tend to be larger than females, to have weapons that females do not, and to be more likely to fight. These characteristics, as Darwin realized, are explainable by the central role of mating in male reproduction. Males evolve large size, weapons, and pugnacity because these traits give them an advantage in competing against other males for mating opportunities.

The other way in which males sometimes differ from females is less easily explained. In some species, and especially in birds, males are ornamented. The male ornaments include bright, conspicuous colors and often extremely exaggerated structures. For example in the marvelous spatuletail hummingbird, *Loddigesia mirabilis*, the male, like most male hummingbirds, has a *gorget*, throat feathers that are specially modified to refract light to produce brilliant, intense color. The male spatuletail is unique, however, in its highly modified tail feathers. The tail contains just four feathers, two of which are extremely long (longer than the bird's body) bare shafts, tipped with circular blue discs.

Ornaments like this cannot be explained as traits that help males to threaten or to hurt each other, and they certainly don't seem designed to help survival, so how do they come about? Darwin thought of an explanation and in doing so displayed his astounding mind as well as his intellectual freedom from the *Zeitgeist*. Darwin proposed that the ornaments evolved to such extremes because they made the males more attractive to females. In other words, the ornaments evolved through the actions of female mate choice. In mid-nineteenth-century Britain, this was

Figure 11.2 Male spatuletail hummingbird. Photograph by Bernardo Roca-Rey Ross.

a bold, radical idea, almost as radical as Darwin's assertion that species are not immutable.

Darwin's idea that the actions of females might count for something was so far ahead of its time that it did not receive any serious thought from biologists until about seventy years later. Even then, biological theoreticians developed only half of the hypothesis. Darwin's hypothesis contains two big ideas. The first is that ornaments, where they exist, might influence male mating success. The second, more important, idea is that the actions of females can determine the outcome of competition among males. Although Darwin, and most of the theoretical biologists who took up his hypothesis, were looking for an evolutionary explanation for male ornaments, the crucial idea, that females might have agency, was in print.

More than a hundred years passed before someone realized the more general significance of Darwin's idea. That person was Robert Trivers, who noted that whenever there is asymmetry in parental expenditure, where one sex spends more time, risk, or energy on offspring than the other, the sex with the greater expenditure may be expected to evolve choice behavior. In most animals, females make the greater expenditure, so they should be the choosy sex. What should they choose? In some instances, it is the male that offers the best resources (for example, territory quality or an actual food gift). In others, it is the male that offers the best genetic material for her offspring. The importance of Trivers's idea is that it predicts where we should expect to see female choice, irrespective of ornaments. Although Darwin was trying to explain ornaments, an ostensible contradiction to his theory of natural selection, he came up with a much bigger idea. It only took biologists 110 years to see it.

We now know that female mate choice, especially in species, as Trivers predicted, where females spend much more on offspring than males do, is common. The mate choice often involves a kind of sampling period, in which the female appears to be comparing potential mates in some way. This is most obvious on leks, where a female can easily move among male territories, and the only purpose of her visit is to find a mate.

In the high sagebrush deserts of North America, male sage grouse defend small territories (no more than about two to six meters in diameter) that are clustered together in clumps of several dozen. At dawn, each male begins to display on his little territory. Females fly in to the vicinity of the lek, and walk among the territories, apparently assessing potential mates. Eventually, the female chooses a male, mates with him, and then flies away, to build a nest and to incubate eggs on her own.

On each territory that she visits, the female is greeted by a male that approaches and displays to her. The principal basis for each male's display is his esophageal sac, which sits in a muscular

Figure 11.3 Male sage grouse at the moment that the strut display produces the poink. Photo by Gail Patricelli.

bag that hangs like a heavy shawl draped across his chest. The sac is an outgrowth of the esophagus and is large, compared to the size of the bird, with an internal volume of four to five liters. To display, a male erects his pointy pail feathers in a fan that frames his head and the esophageal sac with brilliant white feathers. The male raises and lowers the pouch twice in gulping motions that shuttle air from the lungs upward and into the pouch. He produces three soft coos and then, with obvious exertion, contracts the muscular bag around the pouch, pressing it backward into the spine. Then he abruptly relaxes the contraction. The bag swells forward, and two pale featherless patches of skin on the front of the pouch bulge outward briefly. Concurrent with this motion is the culminating sound of the display, described as a "poink." The poink is a composite sound. There are two snaps, produced by the collapsing air sac, spaced at about two-tenths of a second apart. Between the snaps the male uses his syrinx to produce a frequency-modulated (the sound frequency rises, then falls) whistle. Sound energy from the snaps tends to project forward and sound energy from the whistle is distinctly beamed to the right and left. The human ear simply hears poink, but one researcher noted that the effect of the whistle was to make the sound have a more "mellow" quality.

For males, the act of inflating the huge air bags, compressing them violently, and then abruptly releasing the pressure looks taxing. A male performs this display many times in quick succession, especially when a female enters his territory. A male that attracts many females may display at close to a thousand times per day, essentially doubling his daily energy expenditure. Males that display at these high rates lose mass and they are forced, when they take a feeding break off the lek, to travel greater distances to find high quality food.

On a lek, there is a pronounced skew in male mating success. A few males are very popular, and attract and mate with many females, and most males have more limited or no success. When researchers have measured different aspects of males to see what characteristics predict mating success, they found that no physical measurements seemed important. Rather, a particular aspect of the display motion, the time interval between the first and second snaps that make up the poink, predicted male mating success. What information is conveyed in such tiny time differences is unclear, but it seems that females are attending to some aspect of the display that indicates a male's proficiency in motor performance. That is, females are paying attention to male motor skill. In addition, because the displays are so energetically expensive, the females may be measuring, by a male's display rate, his overall vigor.

These two aspects of males, vigor and skill, make sense as traits that a female, looking for good genes for her offspring, should pay attention to. Vigor, the ability to perform energetically expensive acts repeatedly, and skill, the ability to perform physically challenging motor acts well, are likely to be accurate predictors of overall organism performance. Vigor and skill probably reflect an individual male's genetic load, the number of mutations with small deleterious effects that he carries. "Small effect" deleterious mutations are common in animal genomes, and they tend to build up because, individually, they are weakly opposed by natural selection. The fewer the number of small effect deleterious mutations that a male carries, the greater the likelihood that he will be able to perform with the consummate athletic grace that nature demands. A female that can detect subtle differences in male motor performance in effect "peers into" the complete genomes of potential mates. An increasing number of studies are finding that male mating display involves skilled athletic performance, and that females seem to base their mate choice decisions on their evaluations of such performance. It may be that the ornaments that troubled Darwin so much evolved, in the few species that have them, as tricks, secondary enhancers designed to improve the apparent quality of male motor performance.

Male weapons, large body size, aggressive display, and fighting are dramatic and easy to observe. Female choice is less ostentatious; perhaps this is why it was ignored for so long (feminists have another plausible explanation). However, the evolutionary consequences of female mate choice are certainly as comprehensive as the consequences of aggressive competition among males. The full scope of female mate choice remains unknown.

12

Humans

At the moment, you are probably sitting in a building, and you are probably wearing clothes. Those two aspects of technology, buildings and clothes, help to maintain a warm, moist, tropical environment immediately around your mostly hairless skin. You are a tropical creature, and only technology allows you to live outside the tropics.

Humans are apes. We are one species of four that diverged in recent evolutionary time. The other three are gorillas, chimpanzees, and bonobos. We are most closely related to chimpanzees; our DNA sequences are about 99% identical. The last common ancestor between humans and chimpanzees lived about six to seven million years ago, when the populations of apes that would give rise to humans began to come down from life in the trees and specialize on hunting and scavenging meat and gathering plant food in the emerging savannah grasslands of east Africa. Although fossils of terrestrial apes tend to be rare, concentrated paleontological work has produced enough material to provide a good general picture of evolution in the human lineages after the split from the chimpanzee line.

Bipedal locomotion, the most dramatic and pivotal behavioral difference that characterizes humans, appeared quite early. In most physical respects, humans are fairly unimpressive. Compared to our chimpanzee cousins, we are scrawny and weak. However, in one arena – endurance running – our physical performance is quite good. As the ancestors of humans remodeled the pelvis, vertebral column, the knee, and the foot to accommodate bipedal locomotion, they also increased the length of the hind limbs (the legs). Longer hind limbs allow humans to increase running speed, without significantly changing the energy cost of locomotion, by increasing stride length. Many other aspects of the human muscular and skeletal systems are best interpreted as modifications to enhance the efficiency of endurance running. Specialization for endurance running is why humans have scrawny upper bodies and big behinds. A lightweight upper body enhances energy efficiency during running, and the large gluteus maximus muscle, relatively inactive during standing or walking, becomes active during running, to stabilize the tilt of the upper body as it leans forward.

Why did humans evolve to become endurance running specialists? There are two hypotheses. The first is that endurance running evolved as a way to run prey to exhaustion. The second is that it evolved to serve scavenging: picking up bits of carcasses left by other predators. To find carcasses and to get to them quickly, before other scavengers, humans would need to be able to patrol huge areas efficiently. Endurance running would permit this.

Apes in the human line had become efficient bipedal running specialists by about 1.8 million years ago. From that time, until the appearance of anatomically modern humans in Africa at about 150,000 years ago, brain size doubled, by a seemingly steady linear increase over a relatively short time, in evolutionary terms. Many hypotheses to explain the evolution of the extraordinarily large human brain exist, but none offer conclusive evidence. Anatomically modern humans dispersed out of Africa into Europe, across Asia, and into North and South America, about forty thousand years ago. Clear evidence for human agriculture, the domestication and husbandry of plants and animals to provide an abundant, conveniently controlled, steady food supply, does not appear in archeological sites until about ten thousand years before the present. Before that time, throughout the evolutionary history from early hominids to modern humans, all evidence points to a social organization that revolved around small, multifamily, groups that practiced hunting and gathering, defended territories, and were hostile toward other groups. Thus, humans in the present live in an environment that is very different from the one in which we evolved.

To generate hypotheses that explain human perceptions, cognitive skills, and emotional responses to social situations, our starting assumption should be that the particular behavior we want to explain is designed to promote individual fitness in a tribal, hunting and gathering society. Biologists, and some psychologists, only recognized this explicitly about twenty years ago. This new discipline, that seeks to understand human behavior from a biological perspective, is *evolutionary psychology*.

This line of analysis may seem to be in conflict with mental experience, the rich consciousness that we perceive as an omniscient executive that directs everything that we do. However, consciousness is just one of many different functions that the brain performs, and there is no known biological reason to suppose that this particular function would not have evolved in the service of maximizing fitness. Consider behavior that we perform without conscious control. You withdraw your hand from a hot surface before you are consciously aware of the heat. You blink when an object rapidly approaches your face. You breathe. You shiver when you are too cold. Your muscles of facial expression contract to indicate disgust when a partly liquified decomposing rat is placed in front of you. Your eyes almost constantly move in saccades that jitter the fovea from spot to spot. You yawn, sneeze, cough. When you go for a walk, you do not consciously direct the movements of your legs and feet; a walking motor program does that.

Similarly, consciousness does not produce emotions. Anger, sadness, fear, shame, and joy start in brain areas below consciousness. They trigger specific physiological responses that only sometime later are reported to consciousness. The emotion of fear is started when we perceive that we are in danger, and the physiological suite of events known as the stress response is turned on. The stress response is principally orchestrated by the hormones adrenaline and noradrenaline, which are released by the adrenal glands in response to nerve signals from the brain. These hormones increase heart rate and breathing rate, shut down digestion, increase blood pressure: in summary, prepare the body for vigorous muscular effort. Sometime after these events begin, we experience the conscious awareness of fear. The fact that an emotion starts before the consciousness of the emotion indicates that much processing and interpretation of perception goes on in the unconscious brain.

There is now active investigation on the question of whether consciousness initiates any behavior at all. Consider the results of a frequently cited study conducted in the laboratory of the late Benjamin Libet in 1983. The experimental subject sits with his hands resting on a table. In front of him is a clock face on which a single hand makes a full revolution every two and a half seconds. The subject is told to extend his wrist (raise the palm off the table) at any time of his choosing and to note the position of the clock hand when he makes the conscious decision to move. The subject also reports the time at which he perceives that his hand has started to move. Simultaneously, the researchers record electrical activity in the forearm muscles, and with electrodes placed on the scalp, the brain's readiness potential (the electrical activity in the brain's frontal lobes that indicates the initiation of a movement).

The surprising results that Libet reported were that the readiness potential, the brain activity to initiate a movement, starts first. Sometime later, the subject reports the conscious decision to move. A bit later, the subject reports that he has moved. Then a bit later still, the forearm muscles start to contract. The order of events that Libet demonstrated shows that activity in the brain to start a movement precedes the conscious intent to move, and that the sensation that the movement has started occurs before the actual movement. Thus, our behavior, the actual movement that we make, is not immediately dictated by consciousness. Consciousness becomes informed and involved before the actual performance of the motion, but consciousness does not initiate the preparation for the action.

If consciousness is not involved in directing movements, perhaps it has a more executive role in setting goals. However, an increasing body of work shows that goal-setting occurs at an unconscious level also. Recent research challenges the entire notion of conscious will. Biologists and neuroscientists do not yet understand what the role of consciousness is, but they are in increasing agreement on what it is not. It does not seem to be involved in goal setting or in the immediate direction of movements. So, let's continue to consider a few of the findings of evolutionary psychology.

Nausea and Vomiting of Pregnancy

In all cultures, women in the first trimester of pregnancy typically develop what used to be called morning sickness, now more often called pregnancy sickness. They develop heightened sensitivities and aversions to some food and taste odors. Many women who normally relish coffee suddenly find, in early pregnancy, that the smell of coffee is repugnant. The aversion to some foods is so strong that some women feel nauseous and may vomit. This abrupt behavioral shift in food habits, formerly interpreted as an illness, is now understood to be an evolved adaptation. The nausea and vomiting of pregnancy occurs during the first trimester, when the embryo is in the developmental stage of *organogenesis*, in which the basic body plan and the organ systems are assembled. Disruption during organogenesis usually has dire consequences; it may result in the death of the embryo or in profound birth defects. Disruptions later in pregnancy, during the growth phase, after the body plan and organ systems are established, are less likely to cause serious problems.

The odor and taste aversions in the first trimester of pregnancy are typically towards plant parts that contain many secondary compounds. Secondary compounds are the chemical defenses of plants; they evolved to deter herbivores, and have a variety of toxic effects. They may inhibit digestion, poison metabolic pathways, or act as neurotoxins. Many are disruptors of development. Thus, human females evolved a behavioral switch that helps them to avoid eating plants that may contain chemicals that disrupt development of the embryo during the critical period when it is laying out its body plan.

Reproductive Cycles

Human females are unusual mammals in that they do not signal ovulation. In most mammals, the moment of ovulation, when eggs burst from the ovary and enter the top of the reproductive tract, is associated with an event called *estrus*. In estrus, a female mammal broadcasts visual, chemical and behavioral signals of her state, and becomes behaviorally receptive to male courtship. When not in estrus, the female does not tolerate male advances. In monkeys and in other apes, estrus is accompanied by pronounced swelling of the bare red skin around the vulva. This is an obvious, impossible to miss, signal of ovulation. Humans, in contrast, do not have estrus. Ovulation is concealed, even from the ovulating woman, and there is no clearly defined period of sexual receptivity. There is a slight peak in sexual activity and in behavior that could be called sexual solicitation in women around the time of ovulation, but it is only just detectable. In addition, women living in proximity to each other tend to synchronize their menstrual cycles. Synchrony is achieved by subliminal chemical cues in armpit odor.

Why should human females have evolved this specialized form of reproductive cycling? The best guess at present is that the specialization is an outcome of sexual conflict. It is a way to blunt the force of male infanticide. If ovulation is concealed and sexual receptivity is spread across most of the menstrual cycle, a male has very limited information about paternity, not even if his mate has had EPC. Nothing like the infanticide clock that works for male mice is available, because the male has no reliable cue to indicate when the female became pregnant.

A child is born. The primary mate may be the father because he copulated repeatedly with the woman. However, other males may be the sire, because of EPC. There is no easy way for a male to know for certain whether or not he is the father. Thus, the motivation for infanticide is diminished.

This explanation relies on male infanticide being a risk when the male has more certainty of paternity. Is this true? Are human males more likely to practice infanticide when there is certainty that they are not the sire of an infant? We do not know what the ancestral rates of infanticide were, but even in modern western society, where the force of the law operates powerfully, the pattern is clear. When a young infant (between birth and two years of age) is living in a household with the mother and a male who is not the father, the infant's risk of suffering some form of physical abuse from the male is much higher than the risk for a child living in a household with both biological parents. The risk of being killed by the stepfather is also greatly elevated. In Canada, a child under two that lives with the mother and a stepfather has a six in 10,000 chance of being killed by the male. This rate of killing is seventy times greater than the rate at which biological fathers kill their children. The killings of stepchildren seem to stem from moments of rage by the male. In 82% of instances in which a stepfather kills a cohabiting child of under five years of age, they are killed by beating, rather than shooting, suffocating, or other means. A biological parent who kills a child is more likely to use a gun. More recently, researchers who reported the data on infanticides in Canada also summarized evidence to show that the greatly elevated risk of infanticide when a male who is not the biological father is living with a young child is not unique to Canada but is a worldwide phenomenon.

Polygyny

Infanticide, as indicated in Chapter 11, is characteristic of polygynous species. Were humans ancestrally polygynous? There is quite a bit of evidence that they were. First, our species shows sexual size dimorphism. Human males are larger and more muscular than females. Second, there are differences in temperament. Human males are much more prone to physical violence; across many cultures they commit 93–100% of murders. These murders are overwhelmingly about two things: social status striving among other males, and sexual jealousy. Violence in both arenas contributes to mating success.

Third, human males are behaviorally focused on copulation and on maximizing the number of different women that they copulate with. Abundant evidence, summarized by Laura Betzig, shows that with the start of agriculture and the sudden possibility that powerful individuals could monopolize resources, vast asymmetries in male mating success occurred. In culture after culture, a social organization in which a top leader, who controlled a huge harem of females, appeared. Modern worldwide surveys indicate that in every culture, men state that they want many more (about five times more) sex partners than women do, and they indicate willingness to engage in sexual intercourse with an attractive woman almost immediately after meeting her. Women, in contrast, do not reach the male "yes" level until they have known a man for three to six months.

Fourth, human males have an adaptation to deal with sperm competition. If a male has been physically with his partner for some inter-copulation interval (say, two days), on the second copulation he ejaculates about 200 million sperm. If during that interval he has been physically apart from his partner, on the second copulation he ejaculates 600 million sperm. Couples in this study reported times spent apart between copulations that ranged from almost zero to almost 100%, and the number of sperm ejaculated rose predictably as the time spent apart increased. The number of sperm ejaculated had nothing to do with ejaculate volume, and neither member of the couple could perceive the difference. The number of sperm ejaculated is not under conscious control. The male need not even suspect that his partner is having EPC. The adjustment with EPC risk is simply an automatic thing that males do, like sweating when body temperature is too high, or shivering when body temperature is too low.

Do human females, as these results suggest, routinely engage in EPC? Results from a widely distributed anonymous questionnaire suggest that they do. The data showed that as the level of sexual experience (defined by the number of lifetime copulations at the date that the respondent filled out the questionnaire) increased, the percentage of females that reported a double mating (mating with two different males within an hour to a day) rose to about 70%. Almost no females reported being strictly monogamous, and the likelihood of EPC rose as the average time that the male partner was with the female declined. In other words, male mate guarding is needed to prevent EPC.

The principal value of evolutionary psychology is that it has the potential to reveal our true natures. When everyone acknowledges the truth about human nature, social institutions may relax a bit, to be less dissonant with whom we are. In most modern cultures, there is a legal enforcement of monogamy. Evolutionary psychology shows us that strict monogamy is definitely not the natural human mating system. Because of the extensive parental care that human infants require, humans may be more like songbirds. It seems that humans are socially, but not genetically monogamous.

War

There are only two mammal species in which males form cooperative coalitions to defend territory boundaries, kill territorial neighbors, and opportunistically abduct territorial neighbor females for sex. Those two species are chimpanzees and humans. Territorial war occurs in all the modern hunter-gatherer societies that anthropologists have found, and in technological societies it occurs in the context of urban gangs. War is unusual and it is rare among animals, because it is a cooperative activity. Many circumstances work against the evolution of cooperation, and especially against cooperation in which participants risk death. Thus, war may be limited to chimpanzees and humans because these two large-brained species have enough social intelligence to allow individuals to monitor many group members simultaneously, to check for signs of cheating, and to remember who did what. Even when the detection of cheaters is perfect, the only benefit of war that could outweigh the cost (risk of death) is increased mating activity. An increased number of wives, or mates, is the rule for acknowledged warriors in preindustrial societies. Where long-term data are available, the evidence shows that acknowledged warriors have enhanced fitness. They have greater numbers of surviving offspring than non-warriors. In addition, one of the ubiquitous male activities in war is rape, a phenomenon known from antiquity to the present.

Human males have brain reward mechanisms that result in group cohesion and effective coalition formation for war. Consider a passage, written by Karl Marlantes, a combat marine veteran who fought in the Vietnam War. A U.S. platoon was starting to advance toward a fortified hilltop held by the North Vietnamese Army:

> Mellas, watching First Platoon's backs, kept asking in a whisper, "Why? Why? Why?" At the same time, immense excitement

(heathens, infidels) and from the performance of group cere-
monies that excite the reward circuits for coordinated group
activity. It is the basis for clubs, social organizations such as unions,
and for political parties. Finally, the us/them discrimination
allows modern nation states to exist. The fact that political leaders
of entire countries can evoke nationalistic sentiment is testimony
to the ease with which the us/them switch is thrown in humans.

Here is another possible contribution that evolutionary psychology could make. If schoolchildren were taught about human
evolutionary history and the existence of the us/them switch,
they might become more tolerant adults. Rates of violence in
modern technological societies are lower than rates in preindustrial societies. In the past sixty years, in the United States, virulent
forms of racism have been overcome. Our ability to stay on this
path will rest in large part on our willingness to explicitly recognize the reality of the us/them switch, to acknowledge that it is
an evolved human trait that we cannot get rid of.

The us/them switch is the principal source of enmity among
people, and its worst consequence is that it is the basis for genocide. Genocides are recorded from antiquity to the present, and
they always start with us/them discrimination. However, there
are plenty of choleric racial, ethnic and religious hatreds going
on at any time, and most never become genocide. There may be
other behavioral switches that are required to move men to the
point at which they are willing to hack many other people to
death with a machete, but at present, we do not know precisely
what those switches are. We need to identify the antecedent conditions in human evolution that promoted the emergence of the
ability to engage in mass killing. If we could identify key stimuli
that must be present for us/them to be converted into genocide,
we could possibly prevent genocides. Then again, it may turn out
that no such switch can be found. It may be that only the us/
them discrimination, followed by the right opportunity, is
required. Currently, chimpanzees do not commit mass killings of

other chimp groups. They go on organized patrols, in which they probe the defenses of a neighboring territory, and kill the occasional victim when they find him alone. But this may be simply because they lack technology.

Performance

The PET scan device is a way to look at ongoing brain activity. When a neuron is active, sending and receiving hundreds of action potentials per second, it runs its sodium–potassium pumps as fast as possible and consumes a lot of energy. Active neurons thus require the transport services of blood to deliver oxygen and glucose, and to carry away carbon dioxide. PET scans indicate where neurons are active by showing differences in local blood flow across the brain. In a recent study, researchers recorded brain activity while the subject, reclined on her back with her head in the PET scanner, moved her feet on an angled platform, to perform tango steps to music. Even in this easy form of recumbent dancing, the researchers detected over twenty discrete areas, widely distributed in the brain, involved in producing steps that were entrained to music. This PET scan study confirmed what most of us already know: dance is a skilled, difficult activity. A dancer performing a challenging choreography reveals the absolute limits of human motor performance.

Dance occurs in all human cultures, usually in conjunction with music, which provides a rhythmic beat to which dance movements are entrained. Although dance often has ostensible meaning for specific religious or symbolic aspects of culture, it also has at least two deeper functions. First, it is commonly a coordinated group activity. Participation in coordinated group activities is inherently rewarding (often intensely rewarding) for humans. For humans, brain punishment and reward circuits define being alone as bad and define being harmoniously in a group as good. Dancing may thus qualify as a genuinely cooperative activity, in which performers mutually agree to an activity that promotes group solidarity.

The other function of dancing is sexual display. Dancing is another way in which males compete for mating opportunities and in which females can evaluate the possible genetic value of males. Many ethnographic studies have reported that dance functions in courtship, and many have noted that females seem to evaluate male dances based upon the endurance and vigor of the dancer. Modern laboratory studies are arriving at the same conclusion.

Female evaluation of male dance quality is correlated with independent measures of male physical strength. The Dogon is a relatively isolated group of people who have lived along a sandstone escarpment in Mali, about 250 km south of Timbuktu, for several hundred years. They have developed a fascinating cultural institution called masked dancing. Only males dance; individual men wear masks, some quite difficult to manage, but these are only part of the costume. The dancer's entire dress, together with his postures and movements, transform his whole body into a mask.

Beverly Strassmann, who has studied many aspects of the ecology and reproductive lives of these people, notes that some of the masks are dangerous. The "house of eight stories" (see Figure 12.1) mask could break the neck of an unskilled dancer. The stilt-walker, in which the dancer's legs are lashed to tall sticks, is even more dangerous. Although the dancers are formally considered to be anonymous, Strassmann notes that female observers can identify individuals, and she writes that women watch dances "intently." Strassmann has proposed, but not yet tested, the hypothesis that masked dancing is a costly signal of male quality and that the weight and difficulty ratings for different masks might correlate with the dancer's number of wives or offspring.

Evolutionary psychology is a strange discipline, in that the premise, that fundamental human behavior should reflect adaptation to a stone age, tribal, territorial life, seems unassailable, but the actual tests of hypotheses, mainly because of ethical

Figure 12.1 Dogon men dancing in the "house of eight stories" masks. Photograph by Beverley Strassmann.

constraints, often are weak and indirect. Nevertheless, I hope that those who bravely work in this new field will soldier on, and that the established facts from evolutionary psychology will be introduced into school curricula. The most famous of the inscriptions carved into the wall of the Temple of Apollo at Delphi is "know thyself."

Afterword

In less than a century, since the ethologists started work in the 1930s, the study of animal behavior has become firmly established as a branch of biology. Behavioral biologists acknowledge that theirs is an integrative discipline, which calls for synthesis across levels of analysis that span evolution, ecology, anatomy, physiology, and neuroscience. Tinbergen's four questions gave us a useful set of goals to tackle the synthesis, but progress has been uneven. We know most about Tinbergen's second domain – survival value, or function. Biologists have learned how to use mathematical models to specify optimal behavior, and thousands of studies of animals in nature have confirmed that animals usually behave in ways that put them close to the theoretical optima.

We also know enough about Tinbergen's third domain, development, to conclude that much behavior develops through programmed learning – an interaction between instinctive and learned components. Regrettably, this conclusion has not penetrated general awareness; it is still common to hear public discourse about whether a type of behavior is instinctive or learned.

Progress in Tinbergen's domains one (mechanisms) and four (evolution) has been much slower. At present, there are several gaping holes in our knowledge about behavior, including:

1. How do motor patterns evolve? When there is evolutionary change in a motor pattern, how is the change accomplished genetically, and at the level of nervous system organization?
2. What are the indivisible units of behavior? Do units as identified by neuroscientists correspond to units as identified by geneticists?

3. What mechanisms create the moment-to-moment switching of motor output?

4. What are the interactions between motor planning, motor execution, and proprioceptive feedback?

5. By what mechanism do animals identify novelty – a stimulus that they have not encountered before?

6. There is good reason to believe that cognition is modular. In other words, instead of using a kind of all-purpose general intelligence, animals seem to have specific intelligences, such as the ability of Clark's nutcrackers to remember the spatial locations of thousands of cache sites, the ability of some prey species to predict predator behavior, or the ability of individuals, in many species of monkeys and apes, to adjust their response to a group member based upon that member's social rank. How many modules are there? Can we identify predictable links between modules and ecology?

7. How common is evolutionary lag (a trait evolved in the past and is retained in the present, even though the environment has changed) in behavior? Evolutionary lag is the main premise that supports evolutionary psychology, and there are quite a few other examples of lag. However, a firm generalization is missing.

These are just a few of the big unanswered questions that remain in the study of animal behavior. The biological study of animal behavior is a relatively new discipline: Many new and exciting discoveries remain.

Further Reading

1 – The Biology of Behavior

Lorenz, K. (1961) *King Solomon's Ring* Translated by Marjorie Kerr Wilson. Methuen: London.

Tinbergen, N. (1960) *The Herring Gull's World*. Harper & Row: New York.

2 – Multiple Realities

Griffin, D.R. (1958) *Listening in the Dark. The Acoustic Orientation of Bats & Men*. Yale University Press: New Haven.

This now classic book is a fascinating account of the detective story-like way in which Griffin made the discoveries to prove that bats use echolocation.

Catania, K.C. and Remple, F.E. (2004) Tactile foveation in the star-nosed mole. *Brain, Behavior and Evolution* 63: 1–12.

3 – Adjusting Priorities

Davies, N.B., Krebs, J.R. and West, S.A. (2012) *An Introduction to Behavioural Ecology, 4th Edition*. Wiley-Blackwell: Chichester, UK.

4 – Brains and Glands

Eccles, J.C. (1977) *The Understanding of the Brain, Second Edition*. McGrew Hall: New York.

Nelson, R.J. (2005) *An Introduction to Behavioral Endocrinology, 3rd Edition*. Sinauer Associates: Sunderland, MA.

Zupanc, G.K.H. (2010) *Behavioral Neurobiology. An Integrative Approach, 2nd Edition.* Oxford University Press: Oxford, UK.

5 – Instinct

Beecher, M.D. (2010) 'Birdsong and vocal learning during development.' *Encyclopedia of Behavioral Neuroscience* 1, 164–168.

Marler P. (1970) 'Birdsong and speech development: Could there be parallels?' *American Scientist* 58:669–673.

Hailman, J.P. (1969) 'How an instinct is learned.' *Scientific American* 221, 98–108.

6 – Learning

Shettleworth, S.J. (1998) *Cognition, Evolution, and Behavior.* Oxford University Press: New York.

7 – Moving Through Space

von Frisch. K. (1973) *Nobel Lecture.* Available from: www.nobelprize. org/nobel_prizes/medicine/laureates/1973/frisch-lecture.html

Essays on Spatial Navigation. Available from: http://www.pigeon.psy. tufts.edu/asc/toc.htm

Wiltschko, R. and Wiltschko, W. (2009) 'Avian Navigation.' *The Auk*, 126, 717–743.

8 – Genetics

Plomin, R., DeFries, J.C., Knopik, V.S. and Neiderhiser, J.M. (2012) *Behavioral Genetics, 6th Edition.* Worth Publishers.

9 – Living in Groups

Dawkins, R. (1971) *The Extended Phenotype: The Gene as the Unit of Selection.* W.H. Freeman: Oxford, UK.

Byers, J.A. (1997) *American Pronghorn. Social Adaptations and the Ghosts of Predators Past*. University of Chicago Press: Chicago.

DeWaal, F. (1982) *Chimpanzee Politics: Power and Sex among Apes*. Harper & Row: New York.

Prins, H.H.T. (1996) *Ecology and Behaviour of the African Buffalo: Social Inequality and Decision Making*. Chapman & Hall: London.

Hunt, J.H. (2007) *The Evolution of Social Wasps*. Oxford University Press: Oxford, UK.

Schaller, G.B. (1972) *The Serengeti Lion. A Study of Predator-Prey Relations*. University of Chicago Press: Chicago.

10 – Communication

Searcy, W.A. and Nowicki, S. (2005) *The Evolution of Animal Communication: Reliability and Deception in Signaling Systems*. Princeton University Press: Princeton NJ.

11 – Sexual Behavior

Davies, N.B. (1992) *Dunnock Behaviour and Social Evolution*. Oxford University Press: Oxford, UK.

Daly, M. and Wilson, M. (1983) *Sex, Evolution, and Behavior, 2nd Edition*. Willard Grant Press: Boston, MA.

12 – Humans

Wegner, D.M. (2002) *The Illusion of Conscious Will*. MIT Press, Cambridge, MA.

Pinker, S. (2002) *The Blank Slate: The Modern Denial of Human Nature*. Viking Penguin: New York.

Profet, M. (1995) *Pregnancy Sickness: Using Your Body's Natural Defenses to Protect Your Baby-to-Be*. Addison–Wesley: Reading, MA.

Wrangham, R. and Peterson, D. (1996) *Demonic Males: Apes and the Origins of Human Violence.* Houghton Mifflin: New York.

Baker, R.R. and Bellis, M.A. (1995) *Human Sperm Competition: Copulation, Masturbation and Infidelity.* Chapman & Hall: London.

Rossano, M.J. (2003) *Evolutionary Psychology: The Science of Human Behavior and Evolution.* John Wiley & Sons: Hoboken, NJ.

Index